HEALING ORCHIDS

HEALING ORCHIDS

Hong Hai
Nanyang Technological University & The Renhai Clinic, Singapore

Soh Shan Bin
The Renhai Clinic, Singapore

World Scientific

NEW JERSEY · LONDON · SINGAPORE · BEIJING · SHANGHAI · HONG KONG · TAIPEI · CHENNAI · TOKYO

Published by

World Scientific Publishing Co. Pte. Ltd.

5 Toh Tuck Link, Singapore 596224

USA office: 27 Warren Street, Suite 401-402, Hackensack, NJ 07601

UK office: 57 Shelton Street, Covent Garden, London WC2H 9HE

Library of Congress Control Number: 2019945861

British Library Cataloguing-in-Publication Data
A catalogue record for this book is available from the British Library.

HEALING ORCHIDS

ISBN 978-981-120-529-3
ISBN 978-981-120-644-3 (pbk)

For any available supplementary material, please visit
https://www.worldscientific.com/worldscibooks/10.1142/11411#t=suppl

Typeset by Stallion Press
Email: enquiries@stallionpress.com

Contents

Foreword

While many people will likely be familiar with the use of fragrant orchids in cosmetics, or the value of *Vanilla* as a flavouring for food, the use of orchids as medicines may be more surprising. In fact, orchids have been used medicinally by different cultures for centuries.

Chinese tradition says that since the dawn of civilisation, the Chinese people, through the kindness of Sheng Nong, have employed Tian Ma (天麻, *Gastrodia elata*) to treat dizziness, epilepsy, dementia or to assist in recovery from a stroke. Its use has continued in China for more than 2,000 years. Other examples come from the genus *Dendrobium*, whose members are widely used in herbal medicines. For instance, one of our Singapore natives, *Dendrobium crumenatum*, is used to treat cholera, pimples and other maladies in Indonesia.

Healing Orchids presents a compilation of literature research which provides insight into the diversity of orchids used in traditional medicine. Covering seven orchid genera, including an in-depth chapter on *Gastrodia elata*, this book will appeal to orchid enthusiasts and laymen alike. It is interesting and informative, containing not just medical information relating to phytochemistry and allergic reactions, but also botanical details, information on commercial cultivation, and legends and folk tales. It would be a good reference for anyone wanting to know about orchids which have the potential to heal.

I would like to congratulate Professor Hong Hai and Ms Soh Shan Bin for this wonderful publication. I hope that readers will be encouraged to explore the subject further, and continued research will be carried out in the future to ascertain more about the benefits of these wonderful gifts from nature.

Nigel Taylor
Group Director
Singapore Botanic Gardens
National Parks Board

Preface

The Orchid Symposium — From Fundamental Research to Medical Applications was held under the auspices of the Institute for Advanced Studies at Nanyang Technological University, Singapore from 8–9 November 2017. The topics presented covered orchid production, orchid viruses, molecular biology of orchid genes and transcription factors, and therapeutic properties of orchids used in traditional medicine.

It was also an occasion to celebrate the 80th birthday of Professor Hew Choy Sin, an eminent researcher in the plant physiology of orchids who has spent a lifetime studying, teaching and promoting the love and cultivation of orchids. This book is dedicated to Professor Hew.

As the presenters of the Chinese herbal orchid *Dendobrium* at the symposium, we thought it would be useful to compile a small volume starting with chapters derived from papers presented at the symposium that dealt with the fascinating subject of the medicinal uses of orchids. We did further research of the extensive literature on this subject and in total covered a total of seven genera that are commonly used as medicine.

Apart from a chapter contributed by Professor P.C. Leung of the Chinese University of Hong Kong, which reported some pioneering research on the genus *Gastrodia*, the material in this book is based on published research sources. The most important source of reference for us was *Medicinal Orchids of Asia* by Dr Teoh Eng Soon, a treasure house of information compiled in detailed encyclopaedic fashion. Dr Teoh also kindly contributed a fascinating chapter on *Vanda* to this volume.

We have written this book for the general reader and the orchid enthusiast. We hope that readers will be stimulated by it to learn more about orchids and their potential medicinal applications, and to visit such rich displays as may be found in the National Orchid Garden at the Singapore Botanic Gardens, a UNESCO World Heritage Site.

We would like to thank Prof K K Phua, chairman of World Scientific Publishing and Ms Joy Quek, editor of this volume, for their generous encouragement and support, and director Dr Nigel Taylor of the Singapore Botanic Gardens for providing information on some of the species grown in the National Orchid Garden.

Hong Hai
Soh Shan Bin
May 2019

Preface

Orchids are known for her beauty, but few people realize they can also be used as a herb. "Healing Orchids", written by Professor Hong Hai and Ms Soh Shan Bin, provides readers with useful and valuable information on the history, development and potential usage of orchids as herbal medicines. It profiles the medicinal properties of seven well known genera of orchids. The book covers history of how these orchids are used as well as the biomedical properties and medicinal values of the phytochemical compounds from the plants.

The book is an enjoyable read. It is well written with clear illustrations and beautiful photographs of the orchids. It is commendable that the authors have gone to great lengths to ensure the book is an easy and interesting read. The valuable insights on the latest advancement of orchid herbs would be of interest to a wide variety of readers — including teachers, orchid researchers, TCM practitioners, healthcare professionals and the general public.

I would like to express my sincere appreciation and thank you to Hong Hai and Shan Bin for dedicating this book to me. I am truly honoured and humbled by the kind gesture.

Hew Choy Sin

Chapter 1

Orchids in folklore and medicine

Orchids are the largest family of blooming flowers with over 25,000 species and approximately 900 genera, many of them cultivated as house plants. They are somewhat difficult to grow owing to their need for filtered light and high humidity, but wild orchids grow worldwide and can be found in every continent except Antarctica, thriving mainly in tropical and sub-tropical climates.

Orchid plants and flowers range in size and shape. Many grow in tropical forests, producing delicate blooms in a wide array of colours. They vary in size from a few inches to the towering vines of the *Vanilla* orchid.

These beautiful exotic plants have long been associated with romance, love and sex. In fact, the name Orchids (*Orchidaceae* family) derives from the Greek word *orchis*, meaning testicle, which their fleshy underground tubers were thought to resemble.

Various cultures around the world used orchids in sexual and fertility rituals. The Turks, for example, made ice cream from orchid tubers to enhance male performance and in Africa, men used *Ansellia* leaves on their wedding day to produce male children. In various rituals in Asia, women used *Dendrobium* to increase fertility.

In the Middle Ages, Europeans were recorded to have seen slipper orchids sprout from the ground where animals had mated. There is also the exotic story of the Filipino Queen who climbed a tree waiting for her husband to return from a battle. She transformed herself into *Vanda coerulea,* whose flower pattern matched her pale blue gown.[1]

In more delicate settings, orchids have been thought to symbolise desirable attributes. These include love, beauty, fertility, refinement, thoughtfulness and charm.

---

[1] Orchids in folklore and mythology, August 11, 2015. https://www.houzz.com/discussions/3291692/orchids-in-folklore-and-mythology. Retrieved 2 Feb 2019.

They come in every colour except true blue (a tinted blue variety exists). Red orchids symbolise passion, white orchids purity, yellow orchids joy and green orchids good fortune.[2]

1.1 Medicinal Orchids

With their ancient association with sex and fertility, it should not be surprising that orchids have found their way into traditional medicine pharmacopeia, where their therapeutic claims extend far beyond sex and fertility tonics to an enormous range of uses, including neurological conditions, skin diseases, bleeding, coughing, traumatic injuries and cardiovascular diseases.

Most of the medical claims of orchids have not been subjected to modern randomised clinical trials (RCTs), and the few that have rarely turn out positive, which would appear to be surprising considering the large number of alkaloids in orchid tissue.[3] However, like most of traditional medicine, including Ayurvedic and Chinese medicine, there are extensive anecdotal records and case studies that suggest that some of the vaunted therapies may well have a basis in clinical experience, but have escaped validation owing to meth-

---

[2] The orchid flower, its meanings and symbolism. http://www. flowermeaning.com/orchid-flower-meaning/ Retrieved 2 Feb 2019.

[3] Bulpitt, C.J. (2005).

odological limitations of RCTs and the lack of interest in conducting well-designed observational studies.

Some healing herbs may be found in National Orchid Garden, Singapore (Photo credit: Soh Shan Bin).

Seven genera of orchids have been selected to illustrate the rich diversity of medicinal applications of orchids.

One of them, *Gastrodia elata*, has been subjected to more rigorous study at the Chinese University of Hong Kong. Some of the findings and analyses from a biomedical standpoint are presented here to give the reader a flavour of the kind of scientific studies that can and have been done on therapy with orchids.

Unfortunately, not enough of such studies have been conducted. Traditional users of medicines derived from orchids have, for the time being, to be content with placing their faith in the experience of traditional physicians, past and present, as well as diverse anecdotal records of their healing properties.

The rest of this volume presents medicinal orchids for the general public and orchid enthusiast. The seven orchid genera covered, which include dozens of species within them, offer a good representation of the very diverse medicinal use of orchids. Among these genera, *Dendrobium, Gastrodia* and *Bletilla striata* (*Baiji*) have been used by Chinese herbalists for thousands of years. Others like *Habenaria* and *Cypripedium* enjoy enduring applications in Ayurvedic medicine.

Selection of Orchids and Their Principal Medicinal Uses

Orchid Genus	Principal Medicinal Uses
Dendrobium	Treat boils, pimples and other skin problems.
	Sometimes used as a tonic or aphrodisiac.
Gastrodia elata	Treat vertigo, headache and cerebrovascular diseases.
Bletilla	Used for various types of bleeding conditions like gastrointestinal bleeding, coughing blood and hemorrhoids.

(*Continued*)

(Continued)

Orchid Genus	Principal Medicinal Uses
Vanilla	Used as aphrodisiacs as well as to treat syphilis and nervous system disorders like melancholia and hysteria.
Vanda	Considered useful for rheumatism, nervous disorders and inflammation.
Cypripedium	Used to improve blood circulation and to reduce swelling.
Habenaria	Used as a tonic for general body debility, and also as aphrodisiac.

We hope that this little book will help stimulate public interest in orchids and their medicinal uses, and provide readers with a flavour of the potential wonders of healing that may be found in these remarkably colourful and exotic plants, and that more research work may be invested in unearthing the exciting medical potential in them.

Chapter 2

Phytochemical and pharmacological studies

To enable readers to better appreciate the biomedical properties of orchids, we would like to introduce a number of technical terms used in plant biochemistry. Traditional medical systems like Chinese medicine do not use these terms in explaining their medical uses, but instead describe their therapeutic effects using conceptual constructs like *qi*, *jing*, *yin*, *yang*, "dampness" and "wind". It is comforting to know that it may be possible in some cases to explain these medicinal properties in biomedical terms.

Some common terms used in phytochemistry, the study of the many chemical compounds that are naturally derived from plants, are briefly explained below. Readers who are already familiar with these terms, or with less interest in biomedical explanations, may skip this chapter.

Metabolites are organic compounds that are starting materials in metabolism processes. They are simple small structures like vitamins and amino acids absorbed by living organisms and can be used to construct more complex molecules or broken down into simpler ones. Intermediary metabolites may be synthesised from other metabolites and often release chemical energy. For example, glucose can be synthesised to form starch or glycogen, and can be broken down during respiration to obtain chemical energy.[4]

In plants, *secondary metabolites* are produced by the plant itself, but they are not required for the survival of the plant. These substances may be involved in various processes such as inducing flowering, fruit set and abscission. In other words, these substances are required for the formation of flowers and fruits, as well as the detachment of the fruit from the plant. They also act as anti-microbials as well as attractants

---

[4] Metabolomics/Metabolites.   https://en.wikibooks.org/wiki/Metabolomics/Metabolites. Retrieved 19 Feb 2019.

or repellents, and have various other effects on the plant itself and on other living organisms. Hence in the study of herbal medicine, these chemical compounds are researched as potential sources of new drugs for treatment of diseases.

With their long and rich tradition of use in healing, orchids are frequently researched for their secondary metabolites. Among these, flavones, anthocyanins, alkaloids, allergens and photoalexins are the most commonly studied.

Alkaloids act on the nervous system. Many alkaloid stimulants such as morphine, cocaine and nicotine are addictive. Alkaloids in orchids fall into two main classes: the pyrrolizidine- and dendrobine-type alkaloids. The genus *Dendrobium* is the richest in alkaloids. The first alkaloid isolated from the orchids was dendrobine. This alkaloid is known to relieve pain, lower blood pressure and augment salivary secretions.[5] More than 14 other dendrobine-type alkaloids have been discovered from various *Dendrobium* species. These alkaloids are useful pharmacologically but must be used with caution, as in sufficient quantity they can be toxic to humans.

Phenanthrenes occur in higher plants, particularly in orchids, and have received much attention for their

---

[5] Hew, C.S. *et al.* (2006).

cytotoxic properties against specific human cancer cells. This anti-tumour effect is an important reason for much research being conducted on them.

Denbinobin, which is isolated from *Dendrobium nobile*, is one of the phenanthrenes known to have quite potent cytotoxic effects *in vitro* and *in vivo*. Phenanthrenes found in other orchid species are found to have anti-allergic, anti-inflammatory, anti-microbial, anti-oxidant and anti-thrombotic properties. It has been reported that several chemical compounds with anti-microbial properties have been isolated from orchid species *Bletilla striata*.

Phytoalexins, which are bacteriostatic and fungistatic, inhibit the growth of bacteria and fungi. They are usually produced in small amounts in plants and do not cause problems when consumed by humans unless taken in excess. Good agricultural practice does help to reduce the amount of photoalexins found in cultivated orchids.

Stilbenoid, a plant-based chemical compound, has biological effects like anti-inflammation, cancer prevention and protection of heart muscles. *Resveratrol*, a well-known stillbenoid, is also a phytoalexin. It is present on the surface of grapes and is known to help ward off fungi and bacteria on fruits. Claims have been made that it has beneficial effects on plants and animals, such as promoting cell regeneration, but it is also thought to have cytotoxic effects.

Other chemical compounds such as *gigantol* and *moscatilin* have been isolated from *Dendrobium nobile* and found to have anti-mutagenic properties. The chemical compounds found in *Gastrodia elata*, mainly gastrodin and vanillin, possesses anti-inflammatory, anti-coagulation, anti-angiogenic and anti-convulsive properties. Gastrodin also has sedative actions on mice and humans, and might be effective medicine for conditions like anxiety, insomnia, neurasthenia and mental hyper-excitation. Research has shown that gastrodin and vanillin have similar pharmacological mechanisms to that of DOPA drugs used for Parkinson's disease.

Chapter 3

Dendrobium

3.1 Introduction

Dendrobium Sw. is a large genus in the orchid family with approximately 1200 species. Sw. is the abbreviation for Olof Swartz who established this genus in 1799.[6] He was the first specialist of orchid taxonomy and has written a critical review of orchid literature. He also discovered that most orchids have one stamen but slipper orchids have two.

---

[6]Dendrobium. https://en.wikipedia.org/wiki/Dendrobium. Retrieved 23 Jun 2018.

The name of *Dendrobium* came from Greek *Dendron* (tree) and *Bios* (life) which means "one who lives on trees" and refers to the epiphyte. The Chinese name *Dendrobium* as *Shihu*, which means "living on rocks". These two *Dendrobium* species, *D. catenatum* (*D. officinale*) and *D. moniliforme* (syn. *D. candidum*) do grow among rocks. They were recorded in the earliest extant compilation of herbs by the Chinese, known as *Shen Nong Ben Cao Jing* (神农本草经) or *Ben Jing* (本经).

3.2 Special Features of the Orchid

Most medicinal varieties of *Dendrobium* have golden, pink and white flowers. It is sometimes not easy to differentiate among the different species.[7]

In the famous orchid garden within the world-heritage Singapore Botanical Garden, one can find many plants of the *Dendrobium* species, like *Dendrobium chrysotoxum* Lindl.*, Dendrobium crepidatum, Dendrobium fimbriatum, Dendrobium loddigesii, Dendrobium nobile* Lindl., *Dendrobium officinale, Dendrobium pulchellum, Dendrobium wilsonii* and *Flickingeria fimbriata* (*D. plicatile*).

An example of a *Dendrobium* species used commonly in traditional Chinese medicine is the

---

[7] Teoh, E.S. (2016), pp. 258.

Dendrobium chrysotoxum Lindl. characterised by beautiful golden fragrant flowers, each about 4–5 cm in diameter. Twenty or more of them would be on an arched inflorescence in a neatly-arranged manner. The sepals and petals are bright yellow with a round, fringed, hirsute, orange coloured lip.

Dendrobium chrysotoxum Lindl.
(Photo credit: Ratchanee Sawasdijira).

Dendrobium officinale, also known as *Dendrobium catenatum* Lindl., is a species that is extensively used by the Chinese as medicinal herb *Shihu*. The stems are erect, long (9–35 cm), cylindrical, slender (2–3 mm thick), with 3–5 leaves on the upper nodes. Its roots are very fine compared with those of other *Dendrobium* species. Inflorescences are 2–4 cm long with 2–3 flowers. Flowers are star-shaped, about 1.5 cm across, with sepals and petals that are white or yellowish green. The lip is pointed with a reddish blotch and the flower gives off a mild or strong fragrance.

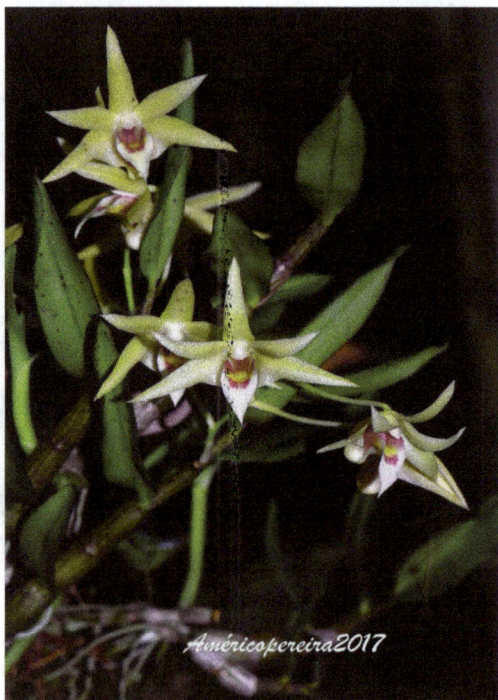

Dendrobium officinale (Photo credit: Américo Pereira. Source: OrchidRoots, http://bluenanta.com/orchid/).

Dendrobium nobile is the current most dominant species of *Shihu*. Besides China, Thailand, India, Nepal, Vietnam and Japan also use it as a medicine.

The pseudobulbs of *Dendrobium nobile* is yellowish green and can be up to 1.2 m (4 feet) tall. In late spring, its pseudobulb loses its leaves and is replaced with an inflorescence of flowers. Each *Dendrobium nobile* Lindl. has stems that bear several inflorescences near their terminals. Each inflorescence carries 2–4 flowers that vary in colour from white, pink to purple, or pink suffused with rosy purple. The flowers are 5.5–8 cm across with a lip that is usually a deep purple blotch in the throat.

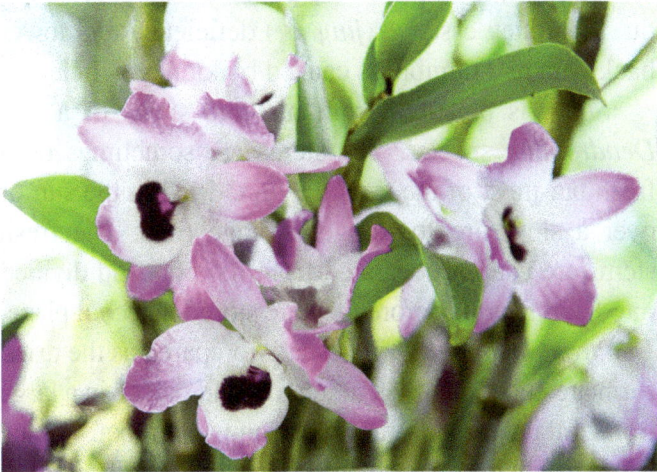

Dendrobium nobile (Photo credit: By Jindowin).

3.3 Traditional Medical Uses

Shihu is a popular tonic which has been used by the Chinese for thousands of years. It is classified as "sweet" in flavour, slightly cold in nature, and attributable to the lung, stomach and kidney meridians. It is a "restorative" (补益药) with the actions of nourishing the stomach, regenerating body fluids, moistening *yin* and clearing heat (养胃生津，滋阴清热). This is the basis for its clinical applications in febrile disease and *xiaoke* 消渴 (akin to diabetes) involving excessive thirst. Abdominal pain, hiccups, vomit, dry cough, night sweat, low grade fever and thirst due to *yin* or fluid deficiency can be resolved by an *yin* tonic like *Shihu*. It is also used for patients with poor vision due to kidney essence (*jing* 精) deficiency and lower back pain and numbness caused by kidney deficiency.

Shihu is prepared from matured stems of a few *Dendrobium* species. The stems are sometimes used fresh but most of the time they are dried in the sun or over a slow fire for longer storage time. The fresh stems are usually separated from its roots and divided into 8 cm lengths before they are stored in a cool place. *Dendrobium moniliforme* usually prepared as the dried form are cut into 4 or 8 cm lengths and placed in a large iron container over a fire to dry and soften before they are twisted into a spiral form and allowed to cool. These are known as the *Er Huan Shihu* (耳环石斛).[8]

Various *Dendrobium* species are used also in Ayurvedic and other traditional medicines. These

---

[8] Teoh, E.S. (2016), pp. 53.

include five *Dendrobium* species, *D. cumulatum*, *D. draconis*, *D. indivisum*, *D. leonis* and *D. trigonopus*, that are used in Thai native medicines.[9] Depending on the source of literature, the Thai word used to describe *Dendrobium* has two different spellings, namely *Uang* and *Ueang*.

Dendrobium cumulatum stems are used to treat asthma, whereas the stems of *Dendrobium draconis* and *Dendrobium trigonopus* are used as anti-pyretic for reducing fevers and haematinic for increasing haemoglobin in the blood through the stimulation of red blood cell production. *Dendrobium indivisum* and *Dendrobium leonis* are used to treat headache in Thailand.

Dendrobium crumenatum is used in the Malay Archipelago to treat ear ache[10] and was the sole orchid listed among 194 species by the Administration Department of the Japanese Army in Singapore. It appeared in their compilation of useful plants in July 1944 (*Compilation of Medicinal Plants in the Malay District*). Malays do not differentiate between species distinguished by botanists and may use dissimilar species for the same medicinal purposes.[11] *D. crumenatum* was the common *Dendrobium* used for poultices (medicinal patches). However, *D. purpureum* was commonly

---

[9] Hew, C.S. *et al.* (2006).

[10] Lawler, L.J. (1986).

[11] Burkhill, I.H. (1935).

used instead in East Malaysia and *D. planibulbe* in West Malaysia. Also in Perak, *D. subulatum* was adopted as a substitute.[12]

In Bangladesh, *Dendrobium nobile* stems are useful for treating dryness of tongue and fever during convalescence. It is also used for enhancing longevity, as an aphrodisiac, and as stomachic for improving digestion and promoting appetite, or simply as an analgesic. Its leaf extract and seed powder are very effective for cuts and wounds. Pseudobulb extract is used to cure eye infections and to soothe burns.

In Vietnam, the plant is used in the treatment of pulmonary tuberculosis, general debility, flatulence, dyspepsia, reduced salivation, parched and thirsty mouth, night sweats, fever and anorexia.[13]

Dendrobium has been used by the Chinese in febrile diseases with fever and dry mouth. This is explained by its action of reducing *yin* deficiency and hence clearing heat that arises from it. In a study by Li *et al.*,[14] dendrobine which is the main alkaloid in *Dendrobium nobile* Lindl. was found to have potent anti-viral activities against influenza viruses through inhibiting early stage viral replication. Dendrobine also has mild anti-pyretic and pain-relieving effects.

---

[12] Teoh, E.S. (2016), pp. 259.

[13] Hossain, M.M. (2010).

[14] Li, R. *et al.* (2017).

Dendrobium is often used in Chinese medicine to treat *xiaoke* in partnership with other herbs to reduce *yin* deficiency. Clinical evidence suggests that oxidative stress plays a major role in the pathogenesis of diabetes.[15] Massive production of reactive oxygen species will destroy beta cells of pancreas, decreasing insulin release. In a study by Pan *et al.*, polysaccharides from *D. huoshanense*, *D. officinale* and *D. nobile* were demonstrated to decrease fasting blood glucose, reduce glycosylated serum protein and increase serum insulin in alloxan-induced diabetic mice. Recovery of injured pancreatic islets was also observed. This study suggests that the polysaccharides decrease sugar level through its anti-oxidative potential.

Dendrobium is also used to treat vision problems. There is a famous medication known as *Shihu Yeguang* pill (石斛夜光丸) in which *Dendrobium* is one of the main herbs and TCM physicians use it to tonify kidneys and hence improve vision. The action of the herb could be explained by research which found that *Dendrobium nobile* helps prevent cataract through its anti-oxidative effect.[16]

As a kidney tonic, *Dendrobium* has the effect of strengthening "healthy *qi*", the rough equivalent of

---

[15] Maritim, A.C. *et al.* (2003).

[16] 魏等 (2008).

boosting the immune system. An experimental study has also shown that a few species of *Dendrobium* (*D. loddigesii* Rolfe, *D. nobile* Lindl., *D. officinalis* and *D. caulis*) can enhance immune response of the white blood cells found in the spleen of the mouse. In another clinical study, *D. huoshanense*, also known as *Dendrobium catenatum*, improved signs and symptoms of patients with atopic dermatitis (e.g. lesions, itching of the skin, insomnia) through modifying the immune response.

Dendrobium huoshanese (*Huoshan Shihu*).

3.4 Biomedical Research into *Dendrobium*

Many of the traditional uses of the *Dendrobium* in Thai native medicine or Ayurvedic medicine are not readily explained in biomedical terms and await further clinical research. In contrast, the medicinal *Dendrobium Shihu* used by the Chinese is arguably the most extensively studied orchid species.[17] There are hundreds of publications on *Shihu*. Much of the focus of research has been on its cytotoxic compounds like denbinobin and erianin. Cytotoxic compounds refer to agents that are toxic to cells, which die when treated with these compounds. *Dendrobium* is also studied for compounds like chrysotoxine, which has neuroprotective functions, and its effect on aquaporins, which are proteins on the cell membrane that are responsible for selective water transport across cells.

In Teoh's encyclopaedic book *Medicinal Orchids of Asia*, it is pointed out that denbinobin was isolated from *Dendrobium* in 1995 and reported then to have cytotoxic effects on human cancer cells in test tube cultures. However, no clinical investigations have followed these findings. Its cytotoxic effect was again shown later with human pancreatic cancer cells *in vitro*. This could indicate an interesting opportunity for more research to further develop it for cancer treatment.

In another study, denbinobin was found to suppress certain lung cancer cell growth and prevent small blood vessel formation.[18]

---

[17] Cited or mentioned at presentation of Teoh, E.S. (2017).

[18] Teoh, E.S. (2016), pp. 318–322.

Erianin, another cytotoxic compound found in *Dendrobium*, is present in *Dendrobium chrysotoxum.* It has been shown to slow down the growth of liver and skin cancer cells and destroy human leukemia cells. It has also been suggested that it can inhibit angiogenesis, which is the process of blood vessel formation that plays a key role in the growth of tumours. Cancer cells can give off certain chemical signals to stimulate the formation of new blood vessels, and these new blood vessels supply blood to the tumour and provide the nutrients and oxygen for its growth. Therefore, the inhibition of angiogenesis is often one of the several treatment methods for cancer. Erianin was also found to cause death of cancer cells in a similar way to the chemotherapy drug, placitaxel, which stops cell division.

Certain compounds from *Dendrobium* have been reported to be able to help kill cancer cells when they have become resistant to conventional cancer drugs. Denbinobin is a naturally occurring phenanthraquinone present in various *Dendrobium* species that has been well researched. In combination with Fas-ligand (another anti-cancer agent), it has a synergistic cytotoxic effect on human pancreatic adenocarcinoma cells. This exciting discovery raises the hope that denbinobin can be an adjuvant used in combination therapies that aim to destroy cancer cells which, relying on decoy receptors, evade the host's immune surveillance. If this can be attained, it would represent a significant step forward for chemotherapy.[19]

S.P. Li at the Chinese University of Macau has shown that there is a great variation of small molecules in *Shihu*, used in Chinese medicine.[20] He has compared five species of *Dendrobium* and their macrophage functions *in vitro* and found that the crude polysaccharide of *Dendrobium officinale* collected from Yunnan Province had the strongest immune-modulatory activities.[21] His investigations suggest that the varied effects on the macrophage functions could be due to the difference in the chemical characteristics of crude polysaccharides from the different species of *Dendrobium*. An investigation on *Dendrobium devonianum* has demonstrated that it can promote the immune function of macrophages and thus could be further explored as a health or functional food.[22] Li concluded that all *Dendrobium* species have polysaccharides with similar chemical characteristics and can possibly be used as a quality marker of *Dendrobium*.[23]

---

[19] This observation is adapted from Teoh, E.S. (2016), p. 259. See also Magwere, T. (2009); Yang, Y.J. *et al*. (2009).

[20] Xu, J. *et al*. (2010).

[21] Meng, L.Z. *et al*. (2013).

[22] Deng, Y. *et al*. (2018).

[23] Cited during symposium presentation by Li, S.P. (2017).

There are many age-related problems for which there are no good drug remedies.[24] This includes exocrine degeneration like dry eyes and dry mouth symptoms — conditions which *Dendrobium* may be able to treat. In modern pharmacological studies, *Dendrobium* was found to have immunomodulatory effect, neuroprotective and hepatoprotective activities, as well as anti-tumour and anti-oxidant effects. It was also shown that it helps to promote body fluids in Sjögren's syndrome.

---

[24] Mentioned at symposium presentation by Leung, P.C. (2017).

Chapter 4

Gastrodia elata

by Ping-Chung Leung[25]

4.1 Introduction

The use of orchids as herbal medicine has a long history in China. *Gastrodia* is one of the three orchids listed in the earliest known Chinese *Materia Medica* (*Shennong Bencao Jing*) [circa 100 AD]. Because of its specific medicinal properties, it may

---

[25] Professor Leung is the Director of the State Key Laboratory of Research on Bioactivities and Clinical Applications of Medicinal Plants (The Chinese University of Hong Kong).

be characterized as an orchid root for the treatment of neurological degeneration.

In China, the *Gastrodia elata* tubers are commonly used to treat various nerve-related symptoms, including epilepsy. *Gastrodia elata* Blume (GE) is a chlorophyll-free orchid found in tropical Asia, Japan and China. Similar species are found in tropical and subtropical areas. In folk medicine, many subspecies are employed for treatment, bearing different names, since they come from different places of cultivation; also the wild type of *Gastrodia elata* is rare. However, all subspecies share some structural characteristics in that:

(1) they have stratifications on the skin with little buddings on the root along the circular stratifications,
(2) all have large apical buddings from both of these tubers, and
(3) indented and grossly umbilicus shaped structures are found at the base of the tubers, which are the connecting points of the tuber and the stem.

These alliances in structure indicate that all belong to the same genus and that all *Tianma* sharing these characteristics are being used as the herbal drug. Their flowers (Fig. 1) are brownish in colour, and sepals and petals are united, forming a cylindrical

Fig. 1.　The *Gastrodia elata* flower.

tube about 1.25 cm long. The growth and maturity of this orchid rely on a special relationship with two fungi: *Mycena osmundicola*, which provides nutrients to the seed during germination, and *Armillaria mellea*, which invades the sprouted tuber and provides nutrients and energy.[1]

Matured rhizomes or tubers (Figs. 2 to 4) are oblong, slightly curved and about 10 to 17 cm in length. In Chinese medicinal practice, this orchid is known as *chi jian* 赤箭 ("red arrow") or *Tianma* 天麻 ("heavenly hemp").[2]

Based on scientific studies, the bioactive compounds in GE are gastrodin (GA), 4-hydroxybenzyl alcohol (4-HBA) and vanillyl alcohol (VA), with

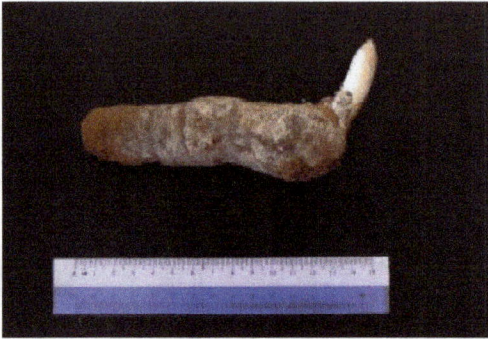

Fig. 2.   Photograph of dried tuber of *Gastrodia elata*.

Fig. 3.   Processed *Gastrodia elata* tuber (*Tianma*).

GA being the major active component. These compounds are pharmacologically active and possess anti-inflammatory, anti-coagulation, anti-angiogenic and anti-convulsive properties.

Fig. 4. Dried *Gastrodia* slices.

4.2 Historical Note — How *Gastrodia elata* Became a Favourite Medicinal Item in the Treatment of Cerebral Symptoms

Searching through the rich collection of traditional Chinese medicine classical records on single herbs and formularies, one could start by going through the ancient terms used to describe symptoms related to the brain, i.e. the cerebral function. Modern terminology for mental disorders and the aging brain of course would not have appeared in ancient Chinese literature. Related terms, however, are plentiful. Chinese terms carrying the meaning of "forgetfulness", "insanity", "change of character" and common symptoms like headache, dizziness, stupor, twitching etc. have been included when herbs are described as possible remedies.[3,4]

Literature review, however, reveals interesting historical discrepancies and group interests. Starting with individual herbs which have been described as indicated for the variety of cerebral symptoms, 87 different items of *Materia medica* can be sourced from 52 different Herb Books (*Ben Cao*) in Chinese. 12 are mineral items, 13 are of animal origin and 56 are medicinal plants. 27 of these plants have been described in more than four of the classic Herb Books as indicated for cerebral symptoms, while the rest are not as popular. The related records have gone through a long period from 220 AD to 1832 AD. The 27 items have since remained popular in different formulations. Interestingly, *Gastrodia elata* does not come under this list.[5,6]

After the review on single herbs (*Ben Cao*), the literature search moves to herbal formulae described in classic records like Herbal Dictionaries of different ages.[7,8] 174 herbal formulae could be extracted from available publications. Within these formulae, a collection of 1676 herbs can be identified. The search could narrow down to those formulae that help with memory loss (taken as an important symptom of dementia), and only 148 herbs are listed. The related formulae have been included in a long historical period from 543 AD to 1900 AD. Again, interestingly, *Gastrodia elata* is not included in this list of herbs commonly used for possibly dementia.[3]

However, within the rich collection of medicinal literature in traditional Chinese medicine, a unique area describing a syndrome related to cerebral vascular accident (stroke) can be located. Thirty-one formulae including 610 individual herbs have been described for this syndrome. About half of the listed herbs are similar to those described under more specific cerebral syndromes as mentioned earlier, while the other half are different. Thus, under this herbal collection commonly used for cerebral vascular accident-related symptoms, *Gastrodia elata* occupied an important position. It is of great interest to note that the control of the clinical symptoms of tremor have also been emphasised.[9,10]

At this point, we may observe that, historically, *Gastrodia elata* had not been a popular choice except in the special area of cerebral vascular accident. Nevertheless, the same unique historical preference might have elicited the special attention of present-day traditional Chinese medicine practitioners who are familiar with cerebral symptoms as well as degenerative syndromes. Indeed, the popular use of *Gastrodia elata* for cerebral vascular accident-related symptoms has prompted traditional Chinese medicine practitioners to include it into their current practice whenever neurological symptoms are found.[11]

May (2009) has done a systematic review of current literature on clinical trials in an effort to understand the results of using Chinese medicine for the treatment of dementia, cognitive impairment and age-related memory impairment. He searched through English language databases which would include the clinical work of active clinicians in China, Korea, Japan and other Chinese communities such as those from Hong Kong. Two hundred and ten articles were considered for detailed evaluation. After initial screening, 125 articles were retained for a Cochrane review.[12]

Randomized controlled trials are most useful references for the level of clinical reliability of the clinical practice. A total of 21 randomized controlled studies had been located in May's report, and a close look at the herbal formulae and herbal items included in the trials would give a quick reference to the popular herbs being used.

In this review of the current practice among traditional Chinese medicine practitioners, *Gastrodia elata* is most frequently included in the prescribed herbal formulations.[3] The popularity of *Gastrodia elata* among clinicians' choice could have influenced the gastronomy experts. Traditional Chinese medicine has always been linked with culinary practice. In South China, using *Gastrodia elata* as a popular component of anti-aging soup and broth has been a common practice.[13]

To conclude, *Gastrodia* has not enjoyed popularity among ancient clinicians prescribing for cerebral symptoms, probably because of its unusual habitat and rarity. However, its remarkable effects on common symptoms like headaches and twitchings have impressed recent-day clinicians and food experts who are giving *Gastrodia* top selection priorities.

4.3 Biomedical Research on *Gastrodia elata*

The main chemical contents in *Gastrodia* are gastrodin (Fig. 5), vanillin and derivatives of vanillin.[1]

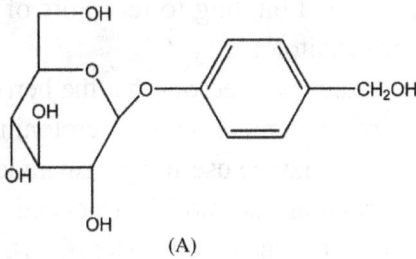

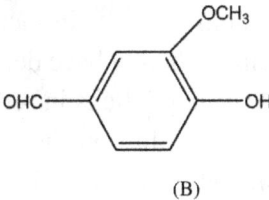

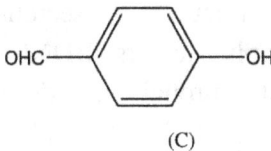

(A)

(B)

(C)

Fig. 5.   Main chemical contents of *Gastrodia elata*.

Gastrodia has been used mainly for ailments of the central nervous system, given orally as a tonic for its neuroprotective activities.[14,15] Early investigations in the 1990s of the last century indicated that two prominent constituents of the plant, *p*-hydroxybenzyl alcohol and vanillin, could exert direct effects on free radicals, thus protecting DNA and amino acids in the central nervous system.[16] Indirectly, the herb could prevent kainic acid binding to receptors of glutamate and thus curb excitotoxicity.[17]

Further studies pointed out that the herb improved the survival of dopaminergic and serotoninergic neurons. This action may be useful against anxiety depression. In depression rat models (induced by forced swimming, tail suspension and open field), *Gastrodia* treatment could improve inhibitory avoidance memory as well as water maze memory.[18,19] In fact, previous studies on *p*-hydroxybenzyl alcohol have demonstrated its effects to attenuate learning deficits in rodent inhibitory avoidance tasks.[20] Vanillin, on the other hand, possessed anti-convulsion effects and might be of value against epilepsy.[21] Upon spinal cord injury, *Gastrodia* promoted the survival of neurons via secretion of more brain-derived neurotrophic factors (BDNF),[15] a protein known to prevent neuronal death during dementia.[22] In other studies, gastrodin glucoside and gastrol identified from phenol extracts all protected H_2O_2-induced PC12 damages in bench works.[23]

In an Alzheimer disease rodent model, *Gastrodia elata* was observed to decrease acetylcholinesterase, thus strengthening acetylcholine activities in the demented brain.[24] Likewise, in a toxic animal model of Parkinsonism induced by 1-methyl-4-phenyl-1,2,3,6-tetrahydropyridine (MPTP), *Gastrodia* extract guarded against apoptosis via regulating the bax/bcl2 ratio, increasing caspases and activating poly (ADP-ribose) polymerase (PARP).[25] In behavioral studies using Drosophila, *Gastrodia* extract prevented amyloid beta-induced damages in locomotion and visual cell deviations.[26] Moreover, in experimental rodents, *Gastrodia* extracts had been documented to block amyloid beta deposits.[27] In another experiment, after bilateral hippocampal injection of amyloid beta of 25–35 amino acids in the rat, treatment of 500–1000 mg/kg of *Gastrodia* for 52 days could lower the need for subsequent increase of acetylcholinesterase in the prefrontal and hippocampal areas. The same treatment also improved the escape latency of the rat in the water maze test.[24] Decrease of acetylcholinesterase would indirectly boost the quantity of acetylcholine in the brain and this has been the logic of therapeutic treatments for senile dementia. Similarly, in another experiment with the rat, scopolamine infused at a dose of 1 mg/kg would lead to memory impairment. This impairment could be prevented by consuming *Gastrodia* at a dose of

500 mg/kg for six weeks.[28] In addition, the anti-inflammatory effects of *Gastrodia* had been studied and the mechanism was related to the inhibition of NF-Kappa B (nuclear factor kappa-light-chain-enhancer of activated B cells) and mitogen-activated protein kinase (MAPK) activation.[29] Furthermore, *in vitro* studies showed that *Gastrodia* could also inhibit expression of inducible nitric oxide synthase (NOS), cyclooxygenase 2 and proinflammatory cytokines in lipopolysaccharides (LPS)-stimulated microglia, via regulation of the MAPK pathway.[30] However, these studies were done only on cell culture platforms or on experimental animals, not humans.

Gastrodia apparently also helps with hypertension affecting brain functions. Experimental studies in the SHR rats (hypertensive rats with a high blood pressure of usually over 210 mmHg systolic starting at young adulthood) with *Gastrodia* treatment of 100 mg/kg, reported the following results: 1) down-regulated the angiotensin I receptors and 2) reduced sera angiotensin II quantity by roughly one-third and aldosterone by roughly one-fifth.[31] Neuronal protection had always been emphasized as a special ability of *Gastrodia* and numerous reports on anti-oxidation, anti-inflammation, blockage of apoptotic cascades and reduction of nitric oxide production/accumulation and energy impairment have been reported (Fig. 6).[32] *Gastrodia elata* appeared to be able to help with the

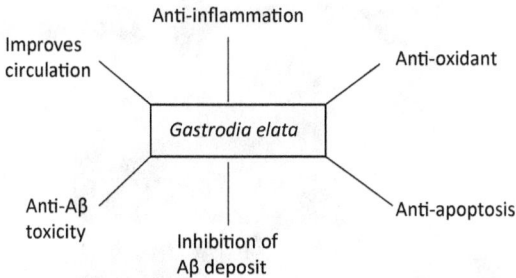

Fig. 6. Multiple Bioactivities of *Gastrodia elata*

reduction of neuronal NOS and prevention of micro-glial activation.[33]

One clinical study on solicited volunteers involving the intake of *Tianma* for two weeks with an oral dose of 15 g a day and comparing with normal age-matched control without medication had been done. Both groups were asked to oppose their thumbs and index fingers for one minute while functional magnetic resonance imaging (fMRI) of blood oxygenation level-dependent (BOLD) images were performed on the brains. Those that had *Gastrodia* displayed more activation sites in their brains and thus reached higher volumes of activation (Figs. 7 and 8). This preliminary result showed that ingestion of the herb did affect the human brain function, either directly as components of the absorbed herb crossing through the blood brain barrier, or indirectly through other smaller metabolites.

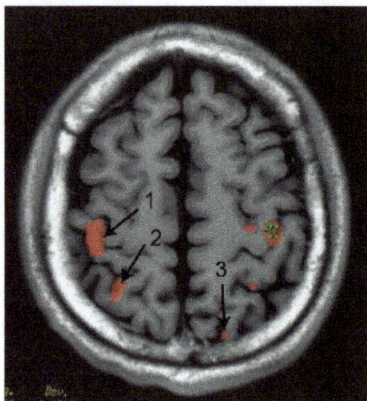

Fig. 7.   Functional MRI image of the normal untreated human while performing thumb and finger opposition. Note activational sites in (1) motor and (2) sensory area while (3) is the visual area. Horizontal section of human brain.

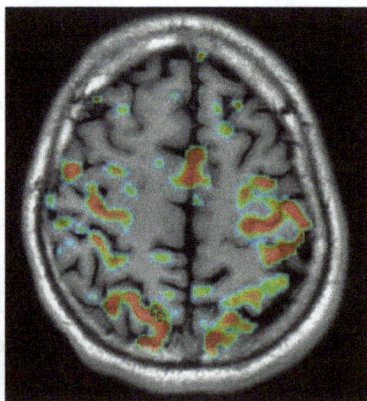

Fig. 8.   Functional MRI image of human on *Tianma* while performing thumb and finger opposition. Note increase of activation sites. Horizontal section of brain.

Experiments had been performed in a rat using Evan Blue extravasation as a means to determine whether the blood brain barrier had been disrupted with *Gastrodia* treatment. This study reported that there were indeed changing levels of Evan Blue extravasation as early as two hours after *Tianma* treatment, signifying that *Gastrodia* or its metabolite had most probably facilitated the blood brain barrier transfer.[34]

In other *in vitro* studies, the ethanol extract of *Gastrodia* was shown to influence human embryonic neural stem cell growth. Subsequent to treatment, there was an increase in the number of spikes on dendritic cells and decreased production of Sox2 and nestin.[35] This might mean better progress towards differentiation.

The anti-convulsive effects of a pharmacological agent could be related to the gamma-aminobutyric acid (GABA) induction.[36] This latter must be carefully monitored as too much GABA induction would lead to depression itself.

Acidic polysaccharide extracts from *Gastrodia* roots suppressed atherosclerotic risk index through inhibition of serum cholesterol composition in the Sprague Dawley rat when fed on a high fat diet.[37] Long-term *Gastrodia* treatment (e.g. for five weeks) in the high fat diet-treated rat reduced the cholesterol by as much as 20%, along with 10% reduction of blood pressure.[38] In a study[39] using mice, cardiac hypertrophy was induced by aortic banding. *Gastrodia* treatment appeared to inhibit cardiac hypertrophy via (1) modulation of ERK1/2 signaling and GATA4 activation and (2) attenuation of fibrosis and collagen synthesis through abrogating the ERK1/2 pathway. Though elimination of the ERK phosphorylation in transgenic mice did not diminish hypertrophic response to pressure overload of ERK1 –/– and ERK2 +/–, the mice showed no reduction in stimulus-induced cardiac growth *in vivo*, but ERK did provide critical protective signals during stress.[40]

A recent study by phytochemists on the mineral contents of *Gastrodia* grown in the different regions of China depicted those genera grown in Tibet had relatively high levels of lead and chromium, probably related to the condition of the soil for cultivation.[41] Herbal medicine is intriguing and useful. A lot of these remedies have been utilized and found useful over the years through trial and error. It is now at a stage of refinement in its quality control and care must be taken in defining the important contents of the herb, which is as important as its therapeutic value and possible toxicity.

In the herb market, many sources of *Gastrodia elata* were dried by sunlight. In some cases, sulfur fumigation was involved in the process. Research employing high performance liquid chromatography fingerprinting revealed there were substantial chemical changes after sulfur fumigation, which therefore should be condemned.[42]

Other pharmacological studies on *Gastrodia* for applications in neurology

There are interesting and important areas of research on *Gastrodia elata* with potential for clinical applications in neurology.

(i) Ischaemic brain damage

Use of *Gastrodia* for the protection of ischaemic damage of brain tissues was investigated using a mouse model after one hour occlusion of bilateral carotid arteries. One group was pre-treated with *Gastrodia*. The fresh brains were obtained 24 hours after reperfusion, then studied for SOD and MDA contents as well as histomorphological changes of the brain tissues. The positive anti-oxidant effects of *Gastrodia* was obvious compared with the untreated group, and the cerebral tissue damages were likewise much less.[43–45]

(ii) Sleep promotion effects of *Gastrodia*

A standard mouse model for the study of barbitone-induced sleep was used to investigate whether *Gastrodia* consumption could shorten the sleep promotion time after intraperitoneal introduction of the sedating agent. The results clearly showed the synergistic effects of barbitone and *Gastrodia*.[46]

(iii) Neuroprotective effects and influence
       of cognitive functions

Neuronal cells were exposed to *Gastrodia* extract or gastrodin which is the marker chemical of *Gastrodia*. The results showed both *Gastrodia* extract as well as gastrodin manifested neuroprotective effects and induced cell proliferations at high concentrations.[47]

The vasodilatory action of *Gastrodia* led to investigations on its specific effects on memory and learning which were related again to dementia. Mouse model and water maze analyses were used. Results showed that *Gastrodia* enhanced cognitive functions and that the mechanisms of action could be an enhanced alpha-secretase mediated proteolytic processing of the amyloid precursor protein, blocking the production of amyloid beta.[48]

4.4 *Gastrodia* Herbal Formulations

Several classical Chinese medical formulations use *Gastrodia* as a principal ingredient.

(i) Da Chuanxiong Fang (大川芎方)

Neurological disorders, like convulsion, stroke and epilepsy, have been treated with *Gastrodia elata* together with rhizome of *Ligusticum chuanxiong* (*Chuanxiong Rhizoma*; CR) in a classical simple formula.

Da Chuanxiong Formula (DCXF) is a famous classical two-herb Chinese medicinal prescription composed of GR and CR (in the ratio of 1:4) for the treatment of headache caused by "blood stasis" and "wind-heat" syndrome. According to the traditional Chinese medicine theory, this formula promotes blood circulation, extinguishes "wind", eliminates pathogens and improves liver function.[49] In terms of modern pharmacology, this herbal pair exhibits significant effects on migraine, neuron protection and cerebral functional improvements. Adding Chuanxiong to *Gastrodia* has the aim of producing synergistic effects to enhance therapeutic efficacy and/or to minimize toxicity and adverse effects. Zhou *et al.* discovered that there are different pharmacological effects of CR and GR in DCXF. CR mainly produces vascular influences while GR works on the neurological side. Synergistic effects and enforced efficiency are probable.[50] DCXF is clinically safe, which has been supported by hundreds of years of popular clinical applications.

To better define the clinical indications of DCXF, relevant experiments have been done. It has been found that phenolic extracts of GR and its active components, vanillin and *p*-hydroxybenzaldehyde, can protect neuron-like cell lines from glutamate-induced toxicity. Water and ethanolic extracts of GE have been proven to have anti-convulsive effects on kainic acid-induced epilepsy in rats. Water extracts of GR were also found to have anti-depressant-like effects on rats. The methanolic extracts of GR and its active components, gastrodin and *p*-hydroxybenzyl alcohol, improved scopolamine-induced memory loss. GR and its active components also possess anti-oxidative and anti-inflammatory functions, which contribute to neuroprotection. CR is widely used as a herbal agent to promote circulation in China. Besides its cardiovascular protective and anti-inflammatory actions, it has excellent anti-oxidative activities to protect cells from DNA damage and apoptosis after ultraviolet B irradiation. The ethanolic and butanolic extracts of CR were able to protect neuron-like cell lines from serum deprivation-induced apoptosis.[16–21] Its active component, tetramethylpyrazine (TMP), was reported as a potent anti-inflammatory and neuroprotective agent. *In vivo* studies indicated that TMP could protect the brain from focal cerebral ischemia and scopolamine-induced neurotoxicity in rats.

Since the creation of DCXF and its wide applications, different Chinese Medicine experts have composed a variety of other formulations basing on the original ratio of the two herbs and the addition of other symptom-relieving herbs for the management of cerebral function-related syndromes.

(ii) Formulations based on neural protection and cerebral vascular improvement

Dementia is a syndrome related to brain pathology, usually of a chronic or progressive nature, associated with neuronal degeneration. Although often understood as primary Alzheimer's disease (~40%) in contrast to vascular dementia (~30%), the two entities are closely related, possibly mixed or it might be vascular deficiency leading to progressive neuronal degeneration. Currently, it is widely understood that the pathological changes could start with focal ischaemia, leading to inflammatory reactions and subsequent abnormal amyloid deposits which hinder neuronal communications. The whole degenerative process is also affected by genetic influences.

With the full knowledge of the complicated pathological background, and the current lack of effective pharmacological measures to prevent or slow down the degenerative processes, the modern way of creating a suitable herbal remedy for dementia would naturally take a dual direction of protecting the neurological tissues from degeneration while at the same time promoting cerebral vascular flow.[51,52]

4.5 Conclusion

Gastrodia elata Blume (*Tianma*) is a remarkable Chinese medicine and its roots have been widely used for the treatment of rheumatism, epilepsy, paralysis, hemiplegia, lumbago, headache and vertigo. The major components present in *G. elata* Blume are gastrodin (GA), vanillyl alcohol (VA) and 4-hydroxybenzaldehyde (HBA). GA is the bioactive component of *G. elata* Blume, which has sedative and anti-convulsant actions, neuroprotective effects, facilitating memory consolidation and retrieval, and anti-oxidant and free radical scavenging activities. Differential methanol extracts of *G. elata* Blume were found to prevent serum-deprived rat pheochromocytoma (PC12) cell apoptosis by the MTT assay and Hoechst staining. Quality, safety and efficacy of herbal medicines remain the important issues to be tackled in the move toward evidence-based medicine. For botanicals and herbal preparations, there is a need to achieve scientific proof and clinical validation with chemical standardization, biological assays, animal models and clinical trials. The analysis of botanicals and herbal preparations will require a holistic solution which involves the combination of the extraction methods and analytical techniques such as separation tools. The major constituents present in *G. elata* Blume had previously been analysed with HPLC and CE methods.

Since *Gastrodia* has a great potential of development in the treatment of neurodegeneration, proper authentication and maintenance of its major chemical components are of utmost importance.

Neurodegeneration and Alzheimer disease, in spite of the decades of research, are yet devoid of adequate treatment. The pathological processes leading to the functional and structural deteriorations are so complex: involving inflammation, oxidation, abnormal protein deposits (tau and amyloid beta), hindering with neural transmissions and normal regenerative processes. Apparently, the standard pharmacological target approach might not work well, particularly when genetic influences are also at play. In the practical search for a medicinal supplement providing balancing effects on multiple targets, a traditional herb like *Gastrodia* could be a good choice. Since it has been found in various studies to have anti-inflammatory, anti-oxidative, anti-amyloid beta activities, as well as shown to have the ability to enhance cell viability and vascular flow, *Gastrodia* deserves further clinical evaluations, either used singly, or in combination with some other herbal components.

While *Gastrodia* is to be kept as the neuroprotective component, the vascular promotion part is to be sought from a pool of Chinese medicinal herbs recognised for having such pharmacological effects. One of the most popular and well-known herbs of this category is *Danshen* 丹参 (*Salvia miltiorrhiza*) which has been in the Chinese market either as individual or in partner compositions. We have made use of the same herb in combination with a less popular herb Puerariae to create a cardiovascular tonic for prevention of atherosclerosis, and have organized evidence-based clinical studies.[53] We decided to use *Danshen* together with *Gastrodia* in an attempt to create a new formula that can protect the neurological tissues from degeneration, while at the same time promote cardiovascular health, possibly through the maintenance of adequate cerebral vascular circulation.

Gastrodia has long been studied for its neuroprotective effects, as has been described earlier. *Salvia* needs to go through *in vitro* tests using amyloid beta-induced toxicity and H_2O_2 — induced oxidative stress assays. Once confirmed on this neuroprotective property for *Salvia*, the optional ratio for *Gastrodia* and *Salvia* when used in combination was worked out using the same tests, and this ratio was found to be *Salvia* 2, *Gastrodia* 1. Subsequently, *in vitro* tests on the inhibition of reactive oxygen species (ROS), amyloid beta-induced apoptosis and acetylcholinesterase (AChE) activities were completed.

In vivo experiments included middle cerebral artery occlusion (MCAO) models which indicated that the combined formula was able to decrease the neurological deficit and cerebral ischaemic infarct volume. Further mechanistic studies demonstrated that the formula could upregulate activities of anti-oxidative enzymes, superoxide dismutase (SOD) and glutathione (GPx). It also downregulated activities of pro-inflammatory cytokines interleukin-6 (IL-6) and tumour necrosis factor alpha (TNFα).

In the 5XFAD transgenic Alzheimer mice model, the formula was found to improve the learning ability and spatial memory of the mice. The improvements were attributed to the inhibition of BACE1 activities, thus diminishing amyloid beta burden, reducing inflammation and supporting glial activities.

The formula was also used in a mouse model of small vessel disease of the brain, using a passive smoking-induced model. The cerebral vessel density, which was reduced after forced smoking, could be improved with the combined formula, presumably due to the suppression of TNFα and IL-6 releases.

All these positive laboratory results have led the way towards the arrangement of proper clinical trials on patients suffering from early signs of neurological deterioration. *Gastrodia*, with its strong historical background and current scientific potential, can be considered a medicinal herb of choice in the battle against dementia.

References*

(1) Ong, E.S., Heng, M.Y., Tan, S.N., Yong, J.W.H., Koh, H.J., Teo, C.C. & Hew, C.S. (2007). Determination of gastrodin and vanillyl alcohol in *Gastrodia elata* Blume by pressurized liquid extraction at room temperature *J. Sep. Sci.* 30: 2130–2137.

(2) Teo, C.C., Tan, S.N., Yong, J.W.H., Hew, C.S. & Ong, E.S. (2008). Evaluation of the extraction efficiency of thermally labile bioactive compounds in *Gastrodia elata* Blume by pressurized hot water extraction and microwave-assisted extraction *J. Chromatogr.* A 1182: 34–40.

(3) May, B. (2009). *Chinese Herbal Medicine for Dementia and Related Disorders*. Thesis Doctor of Philosophy, RMIT University Australia.

(4) Bensky, D. & Clavery, S. (2004). *Chinese Herbal Medicine: Materia Medica*, 35th edition. Eastland Press, Seattle.

(5) Fan, K.W. (2004). Online research databases and journals of Chinese medicine. *J. Altern. Complement. Med.* 10(6): 1223–1228.

(6) Tian, S.S. (2005). Development of a screening system of prescriptions and herbs in the clinical literature of traditional Chinese Medicine. *Zhong Hua Yi Xue Tu Shu Qing Bao Za Zhi*.

---

*This section contains references cited in Chapter 4 only. They are not included in the general Bibliography at the end of this book.

(7) Han, X. (2005). Shu ju wa jue zai zhong yi yao. Zhong yi yao. *Shin Xi* 22(6): 5–6.

(8) Zhou, X.Z., Liu, B.Y. & Wu, Z.H. (2007). Integrative mining of traditional Chinese medicine. Literature and medicine for functional gene network. *Artif. Intell. Med.* 41: 87–104.

(9) Du, G.Y., Zhu, X.C. & Zhao, Y. (2003). Clinical study of effects of tianzhi granules on senile vascular dementia. *Zhong Guo Zhong Yao Za Zhi* 28(1): 73–77.

(10) Wang, L.L., Shen, S.H. & Sui, C.F. (2004). Clinical efficacy of activating blood circulation against stasis in the treatment of vascular dementia. *Shequ Weisheng Baojian* 3(6): 439–441.

(11) Shen, S.L. (ed.) (2006). *Lao Nian Xing Chi Dai Ji Xiang Guan Ji Bing*. Zhong Guo Ke Xue Ji Shu Cha Ban She, Beijing.

(12) Birks, J., Grimley, E.V. & Van Dangen, M. (2002). Gingko biloba for cognitive impairment and dementia. *Cochrane Database Syst. Rev.* 2002(4): CD003120.

(13) Huang, S.S. (2005). *Complete Book of Chinese Recipes for Health Preservation*. Guangdong Tourist Publisher.

(14) Chik, S.C., Or, T.C., Luo, D., Yang, C.L. & Lau, A.S. (2013). Pharmacological effects of active compounds on neurodegenerative disease with gastrodia and uncaria decoction, a commonly used poststroke decoction. *Scientific World Journal* 2013: 896873.

(15) Song, C., Fang, S., Lv, G. & Mei, X. (2013). Gastrodin promotes the secretion of brain-derived neurotrophic

factor in the injured spinal cord. *Neural Regen. Res.* 8(15): 1383–1389.

(16) Liu, J. & Mori, A. (1993). Antioxidant and pro-oxidant activities of p-hydroxybenzyl alcohol and vanillin: Effects on free radicals, brain peroxidation and degradation of benzoate, deoxyribose, amino acids and DNA. *Neuropharmacology* 32(7): 659–669.

(17) Andersson, M., Bergendorff, O., Nielsen, M., Sterner, O., Witt, R. *et al.* (1995) Inhibition of kainic acid binding to glutamate receptors by extracts of gastrodia. *Phytochemistry* 38(4): 835–836.

(18) Zhou, B.H., Li, X.J., Liu, M., Wu, Z. & Ming Hu, X. (2006). Antidepressant-like activity of the *Gastrodia elata* ethanol extract in mice. *Fitoterapia* 77(7–8): 592–594.

(19) Chen, P.J., Liang, K.C., Lin, H.C., Hsieh, C.L., Su, K.P. *et al.* (2011). *Gastrodia elata* bl. attenuated learning deficits induced by forced-swimming stress in the inhibitory avoidance task and morris water maze. *J. Med. Food* 14(6): 610–617.

(20) Wu, C.R., Hsieh, M.T. & Liao, J. (1996). P-hydroxybenzyl alcohol attenuates learning deficits in the inhibitory avoidance task: Involvement of sero-tonergic and dopaminergic systems. *Chin. J. Physiol.* 39(4): 265–273.

(21) Ojemann, L.M., Nelson, W.L., Shin, D.S., Rowe, A.O. & Buchanan, R.A. (2006). Tian ma, an ancient Chinese herb, offers new options for the treatment of epilepsy and other conditions. *Epilepsy Behav.* 8(2): 376–383.

(22) Lorke, D.E., Wai, M.S., Liang, Y. & Yew, D.T. (2010). TUNEL and growth factor expression in the prefrontal cortex of Alzheimer patients over 80 years old. *Int. J. Immunopathol. Pharmacol.* 23(1): 13–23.

(23) Zhang, Z.C., Su, G., Li, J., Wu, H. & Xie, X.D. (2013). Two new neuroprotective phenolic compounds from *Gastrodia elata*. *J. Asian Nat. Prod. Res.* 15(6): 619–623.

(24) Huang, G.B., Zhao, T., Muna, S.S., Jin, H.M., Park, J.I. *et al.* (2013). Therapeutic potential of *Gastrodia elata* blume for the treatment of Alzheimer's disease. *Neural Regen. Res.* 8(12): 1061–1070.

(25) Kumar, H., Kim, I.S., More, S.V., Kim, B.W., Bahk, Y.Y. *et al.* (2013). Gastrodin protects apoptotic dopaminergic neurons in a toxin-induced Parkinson's disease model. *Evid. Based Complement. Altern. Med.* 2013: 514095.

(26) Ng, C.F., Ko, C.H., Koon, C.M., Xian, J.W., Leung, P.C. *et al.* (2013). The aqueous extract of rhizome of *Gastrodia elata* protected drosophila and PC12 cells against beta-amyloid-induced neurotoxicity. *Evid. Based Complement. Altern. Med.* 2013: 516741.

(27) Hu, Y., Li, C. & Shen, W. (2014). Gastrodin alleviates memory deficits and reduces neuropathology in a mouse model of Alzheimer's disease. *Neuropathology* 34(4): 370–377.

(28) Park, Y.M., Lee, B.G., Park, S.H., Oh, H.G., Kang, Y.G. *et al.* (2015). Prolonged oral administration of *Gastrodia elata* extract improves spatial learning and

memory of scopolamine-treated rats. *Lab. Anim. Res.* 31(2): 69–77.

(29) Yang, P., Han, Y., Gui, L., Sun, J., Chen, Y.L. *et al.* (2013). Gastrodin attenuation of the inflammatory response in H9c2 cardiomyocytes involves inhibition of NF-kappaB and MAPKs activation via the phosphatidylinositol 3-kinase signaling. *Biochem. Pharmacol.* 85(8): 1124–1133.

(30) Dai, J.N., Zong, Y., Zhong, L.M., Li, Y.M., Zhang, W. *et al.* (2011). Gastrodin inhibits expression of inducible NO synthase, cyclooxygenase-2 and proinflammatory cytokines in cultured LPS-stimulated microglia via MAPK pathways. *PLoS One* 6(7): e21891.

(31) Liu, W., Wang, L., Yu, J., Asare, P.F. & Zhao, Y.Q. (2015). Gastrodin reduces blood pressure by intervening with RAAS and PPAR in SHRs. *Evid. Based Complement. Altern. Med.* 2015: 828427.

(32) Sun, X.F., Wang, W., Wang, D.Q. & Du, G.Y. (2004). Research progress of neuroprotective mechanisms of *Gastrodia elata* and its preparation. *Zhongguo Zhong Yao Za Zhi* 29(4): 292–295.

(33) Hsieh, C.L., Chen, C.L., Tang, N.Y., Chuang, C.M., Hsieh, C.T. *et al.* (2005). *Gastrodia elata* BL mediates the suppression of nNOS and microglia activation to protect against neuronal damage in kainic acid-treated rats. *Am. J. Chin Med.* 33(4): 599–611.

(34) Wang, Q., Shen, L., Ma, S., Chen, M., Lin, X. *et al.* (2015). Effects of *Ligusticum chuanxiong* and

Gastrodia elata on blood-brain barrier permeability in migraine rats. *Pharmazie* 70(6): 421–426.

(35) Baral, S., Pariyar, R., Yoon, C.S., Kim, D.C., Yun, J.M. *et al.* (2015). Effects of *Gastrodiae rhizoma* on proliferation and differentiation of human embryonic neural stem cells. *Asian Pac. J. Trop. Med.* 8(10): 792–797.

(36) Chen, P.J. & Sheen, L.Y. (2011). *Gastrodiae rhizoma* (Tian Ma): A review of biological activity and anti-depressant mechanisms. *J. Tradit. Complement. Med.* 1(1): 31–40.

(37) Kim, K.J., Lee, O.H., Han, C.K., Kim, Y.C. & Hong, H.D. (2012). Acidic polysaccharide extracts from Gastrodia rhizomes suppress the atherosclerosis risk index through inhibition of the serum cholesterol composition in Sprague Dawley rats fed a high-fat diet. *Int. J. Mol. Sci.* 13(2): 1620–1631.

(38) Lee, O.H., Kim, K.I., Han, C.K., Kim, Y.C. & Hong, H.D. (2012). Effects of acidic polysaccharides from Gastrodia rhizome on systolic blood pressure and serum lipid concentrations in spontaneously hyper-tensive rats fed a high-fat diet. *Int. J. Mol. Sci.* 13(1): 698–709.

(39) Shu, C., Chen, C., Zhang, D.P., Guo, H., Zhou, H. *et al.* (2012). Gastrodin protects against cardiac hypertrophy and fibrosis. *Mol. Cell. Biochem.* 359 (1–2): 9–16.

(40) Kehat, I. & Molkentin, J.D. (2010). Extracellular signal-regulated kinase 1/2 (ERK1/2) signaling in

cardiac hypertrophy. *Ann. N. Y. Acad. Sci.* 1188: 96–102.

(41) Rui, Y., Chen, X., Yu, Y. & Guo, R. (2011). Contents of mineral elements in two traditional Tibetan medicines *Fritillaria ussuriensis* and *Gastrodia elata*. *Arabian J. Chem.*, doi: 10.1016/j.arabjc.2011.08.017.

(42) Zhang, X., Ning, Z., Ji, Chen, Y., Mao, C. *et al.* (2015). Approach based on high-performance liquid chromatography fingerprint coupled with multivariate statistical analysis for the quality evaluation of Gastrodia rhizoma. *J. Sep. Sci.* 38(22): 3825–3831.

(43) Liu, Z.K. (2017). *Neuroprotective Effects of Dachuanxiong Formula in Traumatic Brain Injury*. PhD Thesis, The Chinese University of Hong Kong, Hong Kong.

(44) Lam, P.K., Lo, A.W. & Wong, K.K. (2013). Transplantation of mesenchymal stem cells to the brain by topical application in an experimental traumatic brain injury model. *J. Clin. Neurosci.* 20: 306–309.

(45) Han, C.N., He, F.Y. & Li, Y. (2014). Protective effects of Gastrodia on cerebral ischaemia. Special report from Pharmacology Research Lab, Yunnan 650000.

(46) He, F.Y., Wu, S.Y., Hau, C.N. & Li, Y. (2014). Sedative effects of Gastrodia. Special report from Pharmacology Research Lab, Yunnan 650500.

(47) Ng, C.F., Ko, C.H., Koon, C.M. & Leung, P.C. (2013). Aqueous extract of Gastrodia protected drosophila and PC 12 cells against βamyloid induced

neurotoxicity. *Evid. Based Complement. Altern. Med.* 2013: 516741.

(48) Mishra, M., Huang, J., Lee, Y.Y. & Heese, K. (2011). Gastrodia modules amyloid precursor protein cleavage and cognitive functions in mice. *Biosci. Trends* 5(3): 129–138.

(49) Wang, L., Zhang, R., Qian, Y. & Zhang, Z. (2014). Gastrodin ameliorates depression–like behavior and upgrades proliferation of hippocampal–derived neural stem cells in rats. *Behav. Brain Res.* 266: 153–160.

(50) Zhao, Y.H., Guan, Y. & Wu, W.K. (2016). Potential advantages of a combination of Chinese medicine and bone and marrow mesenchymal stem cell transplantation for removing blood stasis and stimulating neogenesis during ischaemic stroke treatment. *J. Tradit. Chin Med* 32: 289–291.

(51) Zhou, M.M., Yang, K. & Wang, Y.T. (2008). Synergistic effects of dachuanxiong fang in treating migraine. *Pharma. Clin. Chin. Mater. Med.* 24: 6–8.

(52) Zhou, Y. (2003). Thirty-five treatment cases of cerebral arteriosclerosis by Chuanxiong tianma decoction. *Henan Tradit. Chin. Med.* 23: 22–23.

(53) Leung, P.C., Koon, C.M. & Lau, B.S. (2013). Ten years' research on a cardiovascular tonic from quality control and mechanisms of action to clinical trials. *Evid. Based Complement. Altern. Med.* 2013: 319703.

Chapter 5

Bletilla

5.1 Introduction

If you are looking for an orchid that is easy to grow, consider the *Bletilla*. This is a small genus of orchid with six to seven species.[26] One of its common names is "the hardy orchid" because it is able to flourish in both tropical climates and survive winters in temperate countries, although requiring protection from heavy frost and sub-zero temperatures. The Chinese name for *Bletilla* is *Baiji* (白及), which sounds like baiji (白鸡), for white chicken, derived from the resemblance of its

---

[26] Sheehan, M. & Sheehan, T. (2009); Teoh, E.S. (2016), pp. 131.

white fleshy rhizome to a chicken. It is also known as Hyacinth Orchid, *Hyacinth Bletilla*, Urn Orchid, Chinese Ground Orchid, Common *Bletilla* Tuber, Japan orchidee (German), *Mikadoblomma* (Swedish), *Mikadoblomst* (Danish), *Shiran* (Japanese) and *Jaran* (Korean).[27]

Bletilla striata (Photo credit: Todd Boland).

The species of *Bletilla* with medicinal value include *Bletilla foliosa, Bletilla formosana, Bletilla ochracea* and *Bletilla striata*; these species have been used medicinally for more than 2000 years in China.

---

[27] He, X. *et al.* (2017).

The genus *Bletilla* was first described by H.G. Reichenbach in 1852, who found it resembling another genus of orchids known as *Bletia* genus and thus gave it a name that is a diminutive form of *Bletia*.[28]

5.2 Special Features

Bletilla is a sympodial plant which produces new corms and leaves every year. It has underground tuberous rhizomes and a stem that is 20–50 cm tall. Each stem is encircled by three or more folded leaves, and new leaves rise from the corm in the spring. The inflorescences rise from the centre of the new growth and carry 4–6 attractive medium-sized, star-shaped flowers with a lip that has yellow-coloured keels.

It likes well-drained soil and is best with compost and sand. Various species can be found across Asian countries like China, Japan, Taiwan, Vietnam, Thailand and Myanmar.

5.3 Traditional Medical Claims

The *Bletilla* species are not only popular ornamental plants but are often used in Chinese medicine. In particular, the tuber of *Bletilla striata* (Thunb.) Reichb.f. is commonly used to stop several types of bleeding,

---

[28] Sheehan, T. & Sheehan, M. (2009).

including coughing of blood, nose bleeding, bleeding of cracked skin, gastric ulcers or liver cirrhosis leading to vomiting of blood or blood in the stools, haemorrhoids, internal bleeding and traumatic bleeding. It is also used to treat inflammations like ulcers, carbuncles, sores and abscesses. Additional applications include gastrointestinal disorders like flatulence, dyspepsia, dysentery and vomiting.

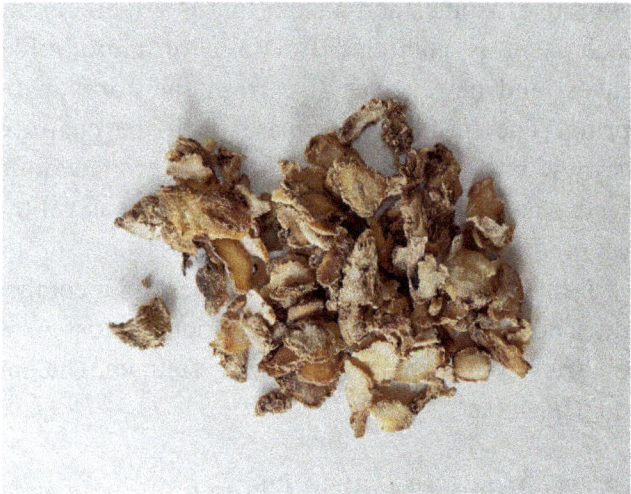

Bletilla striata (Baiji).

In China, Mongolia and Japan, it is used for other purposes like inducing euphoria, purifying blood and treating tumours. It is also used in strengthening and consolidation of the lung and to treat lung conditions like cough, silicosis and tuberculosis. The tubers are also used as demulcent and expectorant, and to treat conditions like fever, chest pain, anthrax, malaria, eye diseases, tinea and ringworm.

The roots are made into powder and mixed with oil to be used as an emollient for skin conditions and burns. Whole plant preparations are sometimes used as tonics and to treat conditions like women's problem of leucorrhoea (white and yellow discharges), coughing of blood and purulent cough. In Japanese folk medicine, the tubers are used for the same purposes as salep, a drink made from tubers of orchid.[29]

B. striata is made into an emollient for the treatment of burns in Vietnam. Pseudobulbs are prescribed for tuberculosis, haemoptysis and other pulmonary diseases. An aqueous extract has been found to reduce bleeding time significantly, as suggested by experiments conducted by Nguyen van Doung.[30] The rhizomes are collected from August to November with a non-metallic tool. In the case of planted crops, harvesting takes place in September and October, 3–4 years after planting. The tubers are crushed or powdered, if dried. They are mixed with oil to produce an ointment useful for treating burns, scalds,

---

[29] Hossain, M.M. (2010).

[30] Nguyen, V.D. (1993).

swellings and cracked skin.[31] The extract of the tubers is an insecticide. *Baiji* can also be used as glue.[32] Essential oils, mucilage and glycogen have been extracted from the tubers.[33]

In Korea, *B. striata* is recorded as odourless and slightly bitter in taste in the Korea Herbal Pharmacopoeia (English edition). It has been widely used as a haemostatic agent in Korean traditional medicine.[34]

Bletilla formosana, known in Chinese as *Xiao Baiji*, is a similar species to *Bletilla striata* and has similar functions of stopping bleeding, treating tuberculous cough, reducing swelling and strengthening lung functions. In India, scrapings of the stem are known to help heal cracked heels.

The other species, *Bletilla foliosa* and *Bletilla ochracea*, are known to have identical traditional medicinal uses as *Bletilla striata*.

5.4 Biomedical Explanations

Bletilla striata polysaccharides (BSP) have been shown to enhance endothelial cell proliferation and promote vascular endothelial growth factors (VEGF),

---

[31] Perry, L.M. *et al.* (1980).

[32] Chen, S.C. *et al.* (1982).

[33] This paragraph has been adapted from Teoh, E.S. (2016), pp. 134.

[34] He, X. *et al.* (2017).

which are essential for formation of new blood vessels during wound healing and tissue repair.[35] This probably explains its use in treating bleeding, ulcers and cuts in traditional medicine.

Chuang Yuling, which contains *Bletilla striata*, is reported to be approved by the Health Ministry of China as a dressing for wounds. The BSP were found to provide scaffoldings.[36]

Anti-microbial agents were isolated from the tubers of *Bletilla striata* in 1983, and some other compounds isolated from the acidic fraction of *Bletilla striata* extract can combat Gram-positive bacteria and are also found to be weakly anti-fungal. Its anti-microbial action probably plays an active role in the healing of wounds.

Baiji has also been used on its own or together with sepium to treat mucosal damage of the bowel and bleeding peptic ulcers. Shi *et al.* (2017) found that phenanthrenes from *Bletilla striata* have strong anti-viral activity for influenza viruses. (Phenanthrenes are compounds that are usually secreted by plants to defend themselves from fungal infection.) This study also suggests that the mechanism for its anti-viral effect derives from its ability to block viral replication.[37] Its anti-viral activity explains the traditional use of *baiji* to treat conditions like fever and cough.

---

[35] Wang, C. *et al.* (2006).

[36] Mentioned by Teoh, E.C. (2017).

[37] Shi, Y. *et al.* (2017).

Bletilla striata has been used as an embolizing agent to treat unresectable liver tumours, and have been used together with modern chemotherapy. It coagulates in blood and blocks the blood supply permanently, and thus is superior to gelfoam powder, which only provides temporary occlusion. A trial in Wuhan Tongji Medical University showed that patients administered with *Bletilla* embolization as compared to conventional gelfoam embolization had a significant decrease in tumour size and reduction in serum alpha-fetoprotein (a marker of liver cancer). The survival rate of patients in the group treated with *Bletilla* embolization was higher than the group treated with gelfoam embolization.[38]

Baiji has been found to be able to kill liver flukes. It is estimated that 35 million people globally are affected by liver fluke, which causes chlonorchiasis. Chlonorchiasis leads to infection of the bile duct and increases the risk of bile duct cancer. Praziquentel (Bayer), which is the only treatment for liver fluke, reduces worm burden by 50–95%, but does not eliminate the infestation. A substance extracted from the *Bletilla* was found to kill the parasite, but no data was produced after that study.[39]

---

[38] Zheng, C. *et al.* (1998).

[39] Reported at the symposium presentation by Teoh, E.S. (2017).

Chapter 6

Vanilla

by Teoh Eng Soon[40]

Ask anyone to name an orchid and most times the answer would be *Aranda*, *Cattleya*, *Dendrobium*, *Phalaenopsis* or *Vanda*. *Vanilla* seldom comes to mind, although *Vanilla* is ubiquitous in the kitchen and its product, vanillin, may be circulating in one's body or clinging to the skin. Stories surrounding

---

[40] Dr. Teoh Eng Soon, MBBS, MD, FRCOG, FACS is a gynecologist in private practice who has researched and written extensively on orchids. His *Medicinal Orchids of Asia* (Springer 2016) is a valuable authoritative reference on medicinal orchids.

Vanilla are filled with love, lust, greed, viciousness and sex, but also generosity and compassion.

Vanilla-flavoured chocolate as a drink was reserved for Aztec nobility who consumed an enormous amount daily, with Monteczuma drinking 50 cups a day (Cameron, 2011). Totonacs and other Indian tribes in Central America offered vanilla as tribute to the Aztecs following their defeat by Itzcoatl (r. 1427–1440).

Monteczuma II offered Cortez a drink of chocolate flavoured with vanilla when he met the Spaniard in November 1519, but the conquistador was only interested in gold. Nevertheless, when Cortez asked Montejoy Portocarrero to convey his Mexican loot to Spain, vanilla was included among the novel products. Spanish nobility failed to appreciate the drink probably because it lacked milk and sugar. "Better thrown to hogs than presented to man," commented a conquistador (Ecott, 2004).

Famous last words! The Spaniard could not foresee that it would become a favourite flavouring item worldwide when Elizabeth I (r. 1558–1603) grew fond of the drink. The Queen's apothecary Hugh Morgan told his queen that vanilla could be added to almost any other item of food or drink. Thereafter, vanilla achieved universal usage in food, beverages, condiments, perfumes, medicines, aromatherapy and even as insect repellents.

Morgan sent vanilla to the illustrious French physician-botanist Carolus Clusius (Charles de l'Ecluse, 1526–1609), who was an economic adviser to the Dutch East India Company (Verenigde Oost-Indische Compagnie, or VOC). Soon, the English monarch was not the only person who had taken to vanilla. Ann of Austria (1601–1666), wife of Louis XIII of France (r. 1610–1643), was fond of vanilla-flavoured chocolate to which sugar and milk had been added and the delightful drink became fashionable with French aristocracy. Madame de Pompadour (Jeanne Antionette Poisson, 1721–1764), the influential mistress of Louis XV, served chocolate flavoured with vanilla and ambergris at her dinners. Cassanova (1725–1798) added vanilla to wine. Marquis de Sade (1740–1814) offered his guests chocolate containing vanilla and Spanish fly that aroused their lustful ardour (Ecott, 2004). This gave birth to the belief that vanilla was an aphrodisiac. In 1847, in an exhaustive book on medicinal botany, an American physician, RE Griffith MD, commented that vanilla was "also considered as acting powerfully on the generative system as an aphrodisiac" (Griffith, 1847).

6.1 Legends of Its Origin

Totonacs, an indigenous people who lived in Santacruz and the southern states of Mexico, were the first people to appreciate vanilla as a flavouring agent. They have their own legends of its origin:

Vanilla originally sprouted from the ground where a beautiful princess, Tzacopontziza (Morning Star), and her lover were slain because their love was taboo due to the fact that she had been consecrated as a celibate priestess to the goddess of fertility. Later, when the priests revisited the site, they discovered a tree with a vine clinging to it. Scattered around on the ground were black pods that emitted a mesmerizing fragrance. The priests concluded that all this came about because of divine intervention. Vanilla pods contained *zanat*, nectar of the gods.

In a second legend, a nobleman blinded by wealth bestowed upon him by the god of happiness betrothed his beautiful daughter Zanath to the lustful god. He did not know that she had secretly fallen in love with a poor artist. When the girl rejected the god, he turned her into a vine. The musician killed himself and became a bee. So it came to pass that every time a flower appeared on the vine, a *melipona* bee would be hovering around it. Their love produced the vanilla capsules (Teoh, 2016).

6.2 Botany

Vanilla Sw. is a pan-tropical genus with more than a hundred species, with 54 occurring in the New World

(Pridgeon, Cribb, Chase, Rassmussen, 1999). Original Aztec. *Vanilla planifolia* has been cultivated in Mexico and the Caribbean for over 500 years. Aztecs call the plant *tlil-xochitl*. In their Nahuatl language *tlil-xochitl* translates as "black flower". Francisco Hernandez de Toledo (1514–1587), who was sent by King Phillip II of Spain to investigate the medicinal plants in the Americas, referred to *Vanilla* as *flore nigro aromatica*. (Hernandez, 1628). Hernandez never saw the plant or the actual flower, which are all green: only the fermented pods are black.

Francisco Hernandez de Toledo (1514–1587).

The first illustration of *Vanilla* was made by Martinus de la Cruz, who authored *Aztec Herball* in 1552. Written in Nahuatl, it was translated into Latin by a fellow collegiate named Juannes Badianus. Thus it came to be known as the *Badanius Manuscript* or the *Barberini Codex,* because it ended up in the library of the Grand Inquisitor. Charles U Clark discovered the manuscript in the Vatican Library in 1929. It should be properly referred to as the *de la Cruz-Badanius Codex*.

The term *"Vanilla"* (*Vaynilla*) was first published in *De Inde utriusque ere naturali et medica* authored by the Dutch physician, Williem Piso (1611–1668) in 1658. Piso had returned from a study tour of Brazil and *De Indiae Utriusque re naturali et medicia* was a 1658 revised second edition of *Historia Naturalis Brasiliae* published in 1648 which did not contain the comment on the orchid. On page 200 of *De Indiae Utriusque* Piso stated: *"est herba Mexicanis Tlilxochitl dicta, quam Franciíc. Hernan- dez Araco Aromatico comparar. Volubilis herba Hederá ... ... quas vulgus Hiispanorum á vaginarum similitudine Vainillas appellar"* [it is a herb which the Mexicans call 'Tlilxochitl', that compares to that called 'Araco Aromatico' by Francisco Hernandez [...] which common Spaniards, because of its similitude with the vagina, call 'Vaynillas'] (Ossenbach, 2017).

Williem Piso (1611–1668).

The generic name *Vanilla* employed by taxonomists was given by Olaf Swartz in 1799.

Orchids belonging to the genus *Vanilla* Sw. are perennial, terrestrial, climbing, epiphytic herbs with branching stems bearing single, fleshy, elliptical leaves at the nodes. Inflorescence is short and axillary, bearing several medium-sized, lemon-yellow or light-green flowers which open in succession and, in general, only lasting the morning. Petals and sepals are free and narrow. Lip is entire or single-lobed, tubular but leaving the column well-exposed. Fruit is large, capsular, shaped like a bean and often referred to as such, but actually not a bean at all and should properly be called capsules (Arditti *et al.*, 2009).

Vanilla seeds germinate on the ground. As the vine develops, it starts to climb the trunk of trees, using axillary roots to hold on. *Vanilla* vines do not twine around tree trunks but climb straight up in a zig-zag manner. Leaves of the Caribbean *Vanilla poitaei* are

Leafless *Vanilla* species (©Teoh Eng Soon, 2019).

reduced to small, stiff "bracts" that assist the stem to latch on as it ascends the tree trunk. An additional six species (*V. aphylla, V. barbellata, V. claviculata, V. dilolniana, V. humboltii, V. madagascariensis, V. phalenopsis, V. roscheri*) are leafless (Cameron, 2011; Gigant *et al.*, 2012). In Asia, *Vanilla aphylla* leaves are reduced to 7 mm long, triangular scales. A central groove runs along the length of leafless stems, which undertake the task of photosynthesis.

Vanilla aphylla (©Teoh Eng Soon, 2019).

Inflorescence is axillary. Flowers are usually borne on drooping side branches. In *Vanilla planifolia* a single flower blooms in succession, but in some species several flowers may open simultaneously on an inflorescence. *Vanilla phalaenopsis,* which is endemic in the Seychelles, has the most attractive flowers in the genus: these are white, 5 cm across, with an apricot patch at the throat, and up to three flowers opening simultaneously (Chase *et al.*, 2017).

Vanilla planifolia (©Teoh Eng Soon, 2019).

In nature, *Vanilla planifolia* is found rooted in soil, in swamp thickets and mixed forests, from sea level to 1200 m (Wiard, 1987). *Vanilla* is hardy and easy to cultivate. Being only a wiry creeper, the potential size of the plants is often not appreciated: it can reach 20 m in height and spread over 2000 m^2 (Soto-Arenas & Dresslers, 2010).

6.3 Fragrant *Vanilla* Species

Vanilla planifolia is the most widely cultivated commercial species and for practical purposes when one discusses *Vanilla*, one is referring to this species. Its fermented capsules are the most fragrant.

Vanilla xtahitensis is a natural hybrid between *V. planifolia* and *V. odorata*. It was brought from the Philippines to Tahiti by Admiral Hamelin in 1848 (Correll, 1953). *V. xtahitensis* is now exclusively cultivated in Tahiti and Moorea for sale to France and Italy, where it is employed mainly to make ice cream. Tahitian *Vanilla* is fruity, smelling of cherry, liquorice, prune and wine. It has a vanillin content of 1.7%. Its unique fragrance is attributed to piperonal and diacetyl (Ehlers *et al.*, 1994).

Vanilla odorata is a fragrant species widely distributed from Mexico across eastern South America to Peru. It is cultivated on a small scale in Mexico for the local market. Cultivation is difficult because the cap-sules are prone to fungus and tend to split before they are ripe for harvest. Mexicans add the vanilla to rum, and formerly it was employed as a vermifuge.

Vanilla pompano is a complex species with sev-eral varieties distributed from Mexico to Peru. Plants and flowers are larger than those of *V. planifolia*. Flowering of *V. pompano* subsp. *grandiflora* in Peru is triggered by a five-degree centigrade fall in ambient temperature. A single vine of subsp. *grandiflora* was once observed to bear 200 flowers. Flowers are pol-linated by bees, *Euleama meriana,* but fruit set is poor. The capsules, commonly known as Platanillo vanilla beans, emit a strong vanilla fragrance but the aromatic profile is different from that of *V. planifolia*. Platanillo vanilla is not an important player in the global market, its sale being mainly confined to Mexico.

Vanilla phaeantha occurs in Florida swamps and bears single, large, ephemeral flowers up to 15 cm across with slim floral segments. The species is resist-ant to disease. It is cultivated in the Antilles as a spice and its fragrant capsules are employed medicinally in Cuba (Henelt, 2001).

Vanilla abundiflora, which is native to Indonesia, was also reported to produce fragrant fruits but they are less fragrant than those of *Vanilla planifolia* (Heyne, 1927). It is not a commercial crop.

Vanilla phaeantha (©Teoh Eng Soon, 2019).

6.4 Medicinal Usage

In the first botanical work of the Americas compiled by a European, Francisco Hernandez made no mention of *Vanilla planifolia* capsules being employed as an aphrodisiac in Mexico but somehow it got to be regarded and employed as such in Europe following its introduction in the 15th century. The infamous Marquis de Sade served chocolate containing vanilla and Spanish fly to provoke 'lustful ardour' among his guests (Ecott, 2004). Yucatan natives employed vanilla capsules 'in local medicine for their supposed excitant and aphrodisiac properties' (Standley, 1930). Several species of *Vanilla*, including *Vanilla planifolia* and *V. madagascarensis*, are employed in decoction in Madagascar with the belief that they are potent aphrodisiacs. In 1847, an American volume on *Medicinal Botany* by R.E. Griffith, MD mentioned that vanilla acts "powerfully on the generative system as an aphrodisiac" and he advocated a dose of 8–10 grains (0.52–0.65 g). Vanilla is still employed as an aphrodisiac in India in the 21st century (Sood *et al.*, 2002).

Vanilla was listed in the *London Pharmacopoeia* in 1721. It was employed to treat hysteria, impotence, gastric ulcers and to assist digestion. However, by the end of the 19th century such usages were obsolete in England (Teoh, 2016).

In India, roots of *Vanilla planifolia* are used as a stimulant and to treat gonorrhoea and dysuria, possibly resulting from the sexually transmitted disease. They are mixed with onions, cumin, sugar and butter to prepare a confection. An extract of the root, together with cumin and sugar, is added to cold milk to provide a remedy for spermatorrhoea (Nadkarni, 1954).

Vanilla (local name *Bejucillo*) was reportedly cultivated for use as perfume by Chocoe Indians in Belize (Duke, 1956).

Vanilla planifolia is not the only species employed medicinally (Teoh, 2016). Other species employed in the Caribbean and elsewhere include:

1. *Vanilla mexicana* (syn. *Vanilla aromatica* Sw.) was employed to treat hypochrondia, hysteria, assist digestion and promote menstrual flow (emmenagogue). Fresh juice of the plant was applied to ulcers, dried stems in syrup employed as vermifuge, and infusion of the roots to treat syphilis in Cuba (Lawler, 1984).

2. *Vanilla claviculata.* The fruit was used in decoction to treat syphilis in Hispanola (Griffith, 1847; Duggal, 1972). In some places it was also used for gonorrhea. Juice from the plant was applied to fresh wounds in Santo Domingo (Lindley, 1838), and the same to promote urine flow and expel worms in Jamaica (Lawler, 1984).

3. *Vanilla poitoei* (syn. *Vanilla eggersii*). Juice from a heated stem was employed as vermifuge in western Cuba (Lawler, 1984).

4. *Vanilla palmarum.* The fruit was used in Brazil to treat disorders of the nervous system (including melancholia), 'asthenic fever and torpor of the uterine system' (Lawler, 1984).

5. Cultivated in the Antilles, *Vanilla phaeantha* is used as a spice, and in Cuba as medicine (Henelt, 2001).

6. *Vanilla decaryi* and *Vanilla madagascarensis* are aboriginal aphrodisiacs in southern Madagascar (Uphof, 1968; Cribb *et al.*, 2009).

7. Irula tribals living at the Nilgiri Biosphere Reserve in India use the stems of *Vanilla walkerie* (local name *Kundu pirandi*) as a veterinary medicine (Balasubramaniam and Prasad, 1996).

8. In Malacca, Alvins (active circa 1884–1888) recorded that flowers of the Malaysian *Vanilla griffithii* were pulped in water for application to the body to relieve violent fever (Burkill, 1935). The large capsules which are fragrant, sweet and tasting like bananas are eaten as fruit (Burkill, 1935; Rifai, 1975; Tanaka, 1976). Sap from the leaves (Uphof, 1968; Duggal, 1972) and stem (Rifai, 1975) is applied to encourage hair to grow.

9. Thai herbalists reported that *Vanilla aphylla* was used to treat liver dysfunction (Chuakul, 2002).

10. *Vanilla wrightii* was employed in western Peninsular India to treat syphilis (Lawler, 1984).

11. George Harley, who worked as a medical missionary in Liberia, described an unusual method of using *Vanilla crenulata* to treat a Manos woman who suffered from dysmenorrhea. Steaming leaves taken from a boiling pot were placed between mats and the woman was

instructed to lie face down on the mat (Harley, 1941; Lawler, 1984). It was also used to treat earache on the Ivory Coast: juice squeezed from heated leaves combined with capsicum was squeezed into the painful ear (Lawler, 1984). This is similar to the way Indian, Malaysian and Indonesian tribals employed the fruit of orchids to treat earache (Teoh, 2016).

Vanilla griffithii (Photo credit: Peter O'Byrne).

Allergic reaction to Vanilla: vanillism

Numerous reports that some vanilla workers handling *Vanilla* capsules suffered from allergic reactions on the face and hands (skin rash, dermatitis and eczema, collectively known as vanillism) have appeared since 1883. It is caused by contact with the whole capsule: pure vanillin evokes a very mild reaction and is not thought to be the cause (Hausen, 1984).

6.5 Phytochemistry

Over 200 compounds have been isolated from vanilla pods viz. vanillin, vanillic acid, *p*-hydroxybenzoic acid, *p*-hydroxybenzaldehyde, proteins, sugars, fiber as hemicellulose and cellulose, waxes, resins, pigments, tannins, minerals and essential oils. Analysis of one lot of vanilla beans gave the following values for the major compounds: vanillin (20%), vanillic acid (1%), *p*-hydroxybenzaldehyde (2%), *p*-hydroxybenzoic acid (0.2%) (Sinha *et al.*, 2007; Rao & Ravishankar, 2000). Vanillin is the predominant compound (20 g/kg) responsible for the characteristic flavor of vanilla, but other volatile constituents (acids, ethers, alcohols, acetals, heterocyclics, phenolics, hydrocarbons, esters and carbonyls) also contribute to the flavor (Klimes & Lamparsky, 1976). Non-volatile components (tannins, polyphenols, amino acids and resins) add to the flavor constituents. Resins assist in the retention of aromatic compounds (Sinha *et al.*, 2007).

Processed vanilla pods (©Teoh Eng Soon, 2019).

Laboratory studies showed that vanillin has anti-microbial and anti-oxidant properties. It prevented mutation of cells exposed to ultraviolet and gamma rays. It lowers lipid levels in Type 2 diabetes (Sinha, Sharma & Sharma, 2007). Some studies showed that vanillin killed cancer cells or prevented their spread (Lirdprapamongkol 2005, 2009, 2010) whereas other studies found that vanillin, at high dosage, may cause cancer cells to multiply and spread (Akagi, Hirose, Hoshiya, *et al*, 1995). An overall assessment of the cancer risk of aldehydes in the diet concluded that vanillin was not a dietary risk factor (Feron *et al*.,

1991). Vanillin was found to potentiate the effect of the cancer drug, doxorubicin, on solid tumours (Eisherbiny *et al.*, 2016). Bromovanin, divanillin and other vanillin-derived compounds also show anti-cancer activity *in vitro* (Yan *et al.*, 2006; Yan *et al.*, 2007; Gomez *et al.*, 2014; Jantaree *et al.*, 2017). These findings need to be investigated by clinical studies before any conclusion can be drawn as to whether consuming vanillin has any benefit on humans, besides enhancing the flavor of food and stimulating the appetite.

Whereas the anti-microbial effects of vanillin suggest that it can be a novel food preservative (Fitzgerald *et al.*, 2004), *Vanilla planifolia* is susceptible to fungal (*Fusarium*) root rot. *Vanilla planifolia* has been bred to disease-resistant species (*V. aphylla; V. phaeantha; V. valsalensis; V. wightiana*) to improve resistance to fungal rot of the roots and to acclimatize *Vanilla* to warmer climates.

6.6 Commercial Cultivation

Vanilla planifolia is by far the most important commercial species, now grown in thousands of hectares in Madagascar, Indonesia, China, Mexico, the Comoros and several African countries. Mexico and the Caribbean, where it was originally cultivated, and

Reunion, where modern commercialization was started, are now not major players in the trade.

When *Vanilla* became popular in Europe, the pods were all imported from Central America. Totonac Indians established *Vanilla* plantations in Vera Cruz in 1767. In 1793 the *Vanilla* plant was taken to the Botanical Gardens in Paris, and then to England. *Vanilla* plants from France were also sent to Reunion island the same year, and further introductions were made during the first quarter of the 19th century. Although the plants flourished, no *Vanilla* was to be had until a 12-year-old Negro slave showed his master, the plantation owner Ferreol Bellier-Beaumont, how to perform hand pollination in 1841. Beaumont did not keep the secret to himself: he taught his neighbours. He also gave the boy his freedom and named him Edmond Albus (1829–1880). Apparently *Vanilla* was hand-pollinated at Liege Botanical Garden in 1836 by Charles Morren and during a lecture in Paris, he showed a vine carrying three fruits (Arditti *et al.*, 2009). Farmers, the people who mattered, were not invited to be privy to the process.

By 1850, more plants were taken from Reunion and Paris to Madagascar, whereupon the crop became an important source of income for the country. Madagascar is now the largest producer of *Vanilla* in the world (Rao & Ravishankar, 2000; Medina *et al.*, 2009).

Edmond Albus (1829–1880).

Most *Vanilla* plantations are located at sea level to 600 m, ideally with slightly sloping ground for good drainage, eastward facing to avoid sun damage of leaves and fruits. The essential requirements are temperature range of 21–32 degrees centigrade, evenly distributed rainfall no less than 1500 mm, and relative humidity of 80%. Soil should have ample organic matter, pH between 6 and 7 and be in excess of 40 cm depth. The aim is to achieve an annual yield of at least 2000 kg of green *Vanilla* capsules per hectare, and five years at full productivity for every orchard (Medina *et al.*, 2009).

Vanilla capsules do not possess the characteristic fragrance when they are harvested. They have to undergo fermentation under controlled conditions. Post-harvest handling involves drying, sweating, curing and additional drying. They are then graded according to size and appearance. Principal buyers grade their purchase by measuring the content and ratios of vanillin, vanillic acid, *p*-hydroxybenzaldehyde and *p*-hydroxybenzoic acid in the capsules. Differences between different origins and batches are commonly encountered (Gassenmeier *et al.*, 2008).

Vanilla farming is bedevilled by a fickle climate. When hurricanes wrecked havoc on plantations in Madagascar, prices of *Vanilla* skyrocketed to US$125 per kg in 2000 for a brief period from a mean value of US$20 per kg in 1997. The 2002 market crash and higher production saw premium pods selling for $20 a kg (Hartman, 2018). Nevertheless, prices have been steadily rising as demand for *Vanilla* continues to climb (Medina *et al.*, 2009), and it was already trading at $500 per kg in 2017 (Gelski, 2017). After the cyclone hit, Madagascar prices soared to $600 per kg (Financial Times, 25 Mar 2018). In the wake of the catastrophe, capsules were harvested prematurely, consequently exacerbating the scarcity of high-quality *Vanilla*.

Korikanthimath *et al.* (1999) proposed that *Vanilla* could be profitably cultivated as a secondary, mixed crop in coconut plantations in India. *Vanilla* is productive in the fourth year after planting, with peak yield at its 6th–8th year. Plants need to be replanted after the tenth year (Medina *et al.*, 2009). Based on a market price for *Vanilla* at Rs 1500/kg (US$22.5/kg), over a four-year productive period, Korikanthimath *et al.* (1999) recorded a net profit of Rs 1500/kg. At current *Vanilla* prices, Indian farmers following his advice would have a windfall. Nevertheless, India is still not among the world's top ten producers of *Vanilla*. The problems Indian farmers faced were described by Sudhashan (2002). China started growing *Vanilla* about 20 years ago and now ranks third among the world's top producers [accounting for 10% of world production in 2006 (Medina *et al.*, 2009)], which is barely sufficient to meet domestic demand.

Vanilla tahitensis is cultivated exclusively on the islands of Tahití and Moorea for sale to France and Italy, where it is employed mainly to make ice cream. *Vanilla pompona* is cultivated exclusively in Guadeloupe and Martinica. It is employed in the manufacture of pharmaceuticals and perfumes.

6.7 Conclusion

The set-up for *Vanilla* production, stringent quality control measures for the product, and maintenance of the gene pool in perpetuity provide valuable case studies for the production of all medicinal herbs.

References[†]

Akagi, K., Hirose, M., Hoshiya, T. *et al*. (1995). Modulating effects of ellagic acid, vanillin and quercetin in a rat medium term multi-organ carcinogenesis model. *Cancer Lett*. 94(1): 113–121.

Arditti, J., Rao, A.N. & Nair, H. (2009). Hand pollination of Vanilla: How many discoverers? In: Kull, T., Arditti, J. & Wong, S.M. (eds.) *Orchid Biology: Reviews and Perspectives X*. Springer.

Balasubramaniam, P. & Orasad, S.N. (1996). Ethnobotany and conservation of medicinal plants by Irulas of Nilgiri Biosphere Reserve. In: Jain, S.K. (ed.) *Ethnobiology in Human Welfare*. Deep Publications, New Delhi.

Bezerra, D.P., Soares, A.K.N. & de Sousa, D.P. (2016). Overview of the role of vanillin on redox status and cancer development. *Oxid. Med. Cell Longev*. 2016: 9734816. PMCID: PMC5204113, Published online 19 Dec 2016, DOI: 10.1155/2016/9734816.

[†] This section contains references cited in Chapter 6 only. They are not included in the general Bibiliography at the end of this book.

Burkill, H.I. (1935). *A Dictionary of Economic Products of the Malay Peninsula*. Crown Agent for the Colonies, London.

Cameron. K. (2011). *Vanilla Orchids: Natural History and Cultivation*. Timber Press, Portland/London.

Chase, M., Christenhusz, M. & Mirenda, T. (2017). *The Book of Orchids*. Ivy Press, London.

Chuakul, W. (2002). Ethnomedical uses of Thai orchidaceous plants. *Mahidol Univ. J. Pharm. Sci.* 29(3–4): 41–43.

Corell, D.S. (1953): Vanilla, its botany, history, cultivation, and economic import. *Economic Botany* 7: 291–358.

Cribb, P.J. & Hermanns, J. (2009). *Field Guide to the Orchids of Madagascar*. Royal Botanic Gardens, Kew.

Duggal, S.C. (1972). Orchids in human affairs. *Acta Phytother.* 19: 163–173.

Duke, J.A. (1956). Ethnobotanical observations of the Chocoe Indians. *Econ. Botany* 10: 280–293.

Ecott, T. (2004). *Vanilla: Travels in Search of the Ice Cream Orchid*. Grove Press, New York.

Ehlers, D., Pfister, M. & Bartholoma, S. (1994). Analysis of Tahiti vanilla by high-performance liquid chromatography. *Z. Lebensm. Unters. Forch.* 199(1): 38–42.

Eisherbiny, N.M., Younis, N.N. & Shaheen M.A. (2016). The synergistic effect between vanillin and doxorubicin in ehrlich ascites carcinoma solid tumor and MCF-7 human breast cancer cell line. *Pathol. Res. Pract.* 212(9): 767–777, DOI: 10.1016/j.prp.2016.06.004.

Feron, V.J., Til, H.P. & de Vrijer, F. (1991). Aldehydes: Occurrence, carcinogenic potential, mechanism of action and risk assessment. *Mutat. Res.* 259(3–4): 363–385, DOI: 10.1016/0165–1218(91)90128–9.

Fitzgerald, D.J., Stratford, M. & Gasson, M.J. (2004). Mode of antimicrobial action of vanillin against *Escherichia coli, Lactobacillus plantarum* and *Listeria innocua. J. Appl. Microbiol.* 97: 104–113, DOI: 10.1111/j.1365–2672.2004.02275.x.

Gassenmeier, K., Reisens, B. & Magyar, B. (2008). Commercial quality and analytical parameters of cured vanilla bean (*Vanilla planifolia*) from different regions from the 2006–2007 crop. *Flavour Fragrance J.* 23(3): 194–201.

Gelski, J. (2017). Madagascar vanilla crop improves, but prices may stay high. *Food Business News,* 12 May 2017.

Gigant, R.L., Brugel, A., de Bruyn, A. *et al.* (2012). Nineteen polymorphic microsatellite markers from two African Vanilla species: Across-species transferability and diversity in a wild population of *V. humblotii* from Mayotte. *Conserv. Genet. Res.* 4(1): 121–128.

Gomez, R.C., Witter, S.K., Hicke, M. *et al.* (2014). Vanillin-derived antiproliferative compounds influence Plk1 activity. *Bioorg. Med. Chem. Lett.* 24(21): 5063–5069, DOI: 10.1016/j.bmcl.2014.09.015.

Griffith, R.E. (1847). *Medicinal Botany.* Lea & Blanchard, Philadelphia, p. 639.

Harley, G.W. (1941). *Native African Medicine with Special Reference to Its Practice in the Mano Tribe of Africa.* Harvard University Press (reprinted).

Hartman, L.R. (2018). Vanilla shortage not as bad as expected. *Food Process.* 17 Jul 2018.

Hausen, B.M. (1984). Toxic and allergic orchids. In: Arditti, J. (ed.) *Orchid Biology III*. Cornell University Press, Ithaca & London.

Henelt, P. (ed.) (2001). *Mansfeld's Encyclopedia of Agricultural and Horticultural Crops*. Springer and IPK, Switzerland.

Hernandez, F. (1628). *Rerum medicarum Novae Hispaniae Thesaurus*, p. 38 (1628) illustrated Vanilla platyfolia [as *Tlilxochitl, seu Flos niger*].

Heyne, K. (1927). *De Nuttige Planten van Nederlandsch Indie*, 2nd (revised and enlarged) edition. Vol. 1. Nijverheld & Handel, Buitenzorg.

Ho, K.L., Chong, P.P., Yazan, L.S. & Ismail, M. (2012). Vanillin differentially affects azoxymethane-injected rat colon carcinogenesis and gene expression. *J. Med. Food* 15(12): 1096–1102, DOI: 10.1089/jmf.2012.2245.

Jantaree, P., Lirdprapamongkol, K., Kaewari, W. *et al.* (2017). Homodimers of vanillin and apocynin decrease the metastatic potential of human cancer cells by inhibiting the FAK/PI3K/Akt signaling pathway. *J. Agric. Food Chem.* 65(11): 2299–2306.

Klimes, I. & Lamparsky, D. (1976). Vanilla volatiles, a comprehensive analysis. *Int. Flavours Food Additive* 7: 272–291.

Korikanthimath, V.S., Hiremath, G.M., Venugopal, M.N. & Rajendra, H. (1999). Feasibility of vanilla cultivation in coconut. *J. Med. Aromatic Plant Sci.* 21: 1033–1039.

Lawler, L.J. (1984). Ethnobotany of the Orchidaceae. In: Arditti, J. (ed.) *Orchid Biology: Reviews and Perspectives III*. Cornell University Press, Ithaca & London.

Lindley, J. (1838). *Flora Medica: A Botanical Account of All the More Important Plants Used in Medicine in Different Parts of the World*. Longman, Orme, Brown, Green & Longmans, London.

Lirdprapamongkol, K., Kramb, J.P., Suthiphongchai, T. *et al.* (2009). Vanillin suppresses metastatic potential of human cancer cells through PI3K inhibition and decreases angiogenesis *in vivo*. *J. Agric. Food Chem.* 57(8): 3055–3063, DOI: 10.1021/jf803366f.

Lirdprapamongkol, K., Sakurai, H., Kawasaki, N. *et al.* (2005). Vanillin suppresses *in vitro* invasion and *in vivo* metastasis of mouse breast cancer cells. *Eur. J. Pharm. Sci.* 25(1): 57–65.

Lirdprapamongkol, K., Sakurai, H., Suzuki, S. *et al.* (2010). Vanillin enhances TRAIL-induced apoptosis in cancer cells through inhibition of NF-kappaB activation. *In Vivo* 24(4): 501–506.

Medina, J.D.L.C., Jiménes, G.C.R. & García, H.S. (2009). Vanilla: Post-harvest operations. *IPhO-Post-Harvest Compendium*. Food and Agriculture Organization (FAO), United Nations, pp. 1–49.

Nadkarni, A.K. (1954): Dr. K.M., Nadkarni's, Materia Medica. Popular Book Depot, Bombay.

Ossenbach, C. (2017). Precursors of the botanical exploration of South America. Wilhelm Piso (1611–1678) and Georg Marcgrave (1610–1644). *Lankerteriana* 17.

Other Fragrant *Vanilla* species.

Pridgeon, A.M., Cribb, P.J., Chase MW and Rasmussen FN (1999): *Genera Orchidacearum Vol.* 2. Oxford University Press, Oxford.

Rao, S.R. & Ravishankar, G.A. (2000). Vanilla flavor: Production by conventional and biotechnological routes. *J. Sci. Agric.* 80: 289–304.

Rifai, M.A. (1975). Extraordinary uses of orchids in Indonesia. *Proceedings of the First ASEAN Orchid Congress.* Kasetsart University, Bangkok.

Sinha, A.K., Sharma, U.K. & Sharma, N. (2007). A comprehensive review on vanilla flavor: Extraction, isolation and quantification of vanillin and other constituents. *Int. J. Food Sci. Nutri.* 59(4): 299–326.

Soto-Arenas, M.S.A. & Dressler, R.L. (2010): A Revision of the Mexican and Central American Species of Vanilla Plumer ex Miller with a characterization of their ITS region of the nuclear ribosomal DNA. *Lankasteriana* 9(3): 285–354.

Sood, S.K., Rana, S. & Lakhanpal, T.N. (2002). *Ethnic Aphrodisiac Plants.* Scientific Publishers, Jodhpur.

Standley, P.C. (1930). *Flora of Yucatan,* Vol. 3, No. 3. Field Museum of Natural History, Chicago, p. 240.

Sudhashan, M.R. (2002). Strategies for increasing production and export of Vanilla. In: *Proceedings of the National Seminar on Strategies for Increasing Production and Export of Spices,* 24–26 Oct 2002, Calicut, Kerala, pp. 18–29.

Tanaka, J. (1976). *Tanaka's Encyclopedia of Edible Plants of the World.* Keigoku, Tokyo.

Teoh, E.S. (2016). *Medicinal Orchids of Asia.* Springer, Switzerland.

Uphof, J.C.T. (1968). *Dictionary of Economic Plants.* Verlag von J Cramer, Lehre.

Wiard, L.A. (1987). *An Introduction to the Orchids of Mexico*. Cornell University Press, Ithaca & London.

Yan, Y.Q., Xu, Q.Z., Wang, L. *et al.* (2006). Vanillin derivative 6-bromine-5-hydroxy-4-methoxybenzaldehyde-elicited apoptosis and G2/M arrest of Jurkat cells proceeds concurrently with DNA-PKcs cleavage and Akt inactivation. *Int. J. Oncol.* 29(5): 1167–1172, DOI: 10.3892/ijo.29.5.1167.

Yan, Y.Q., Zhang, B., Wang, L. *et al.* (2007). Induction of apoptosis and autophagic cell death by the vanillin derivative 6-bromine-5-hydroxy-4-methoxybenzalde-hyde is accompanied by the cleavage of DNA-PKcs and rapid destruction of c-Myc oncoprotein in HepG2 cells. *Cancer Lett.* 252(2): 280–289, DOI: 10.1016/j.canlet.2007.01.007.

Chapter 7

Vanda

7.1 Introduction

Among all orchids, the *Vanda* genus would be the most familiar to residents of the garden city of Singapore, known also as the "City within a Garden". The national flower of Singapore, *Vanda Miss Joaquim*, is an orchid with purplish flowers crossed from two breeds of *Vanda*, *Vanda hookeriana* and *Vanda teres*.

The story of the birth of this hybrid species was controversial until 2016, when the National Parks Board of Singapore concluded that *Vanda Miss Joaquim* was indeed a hybrid crossed from its parent

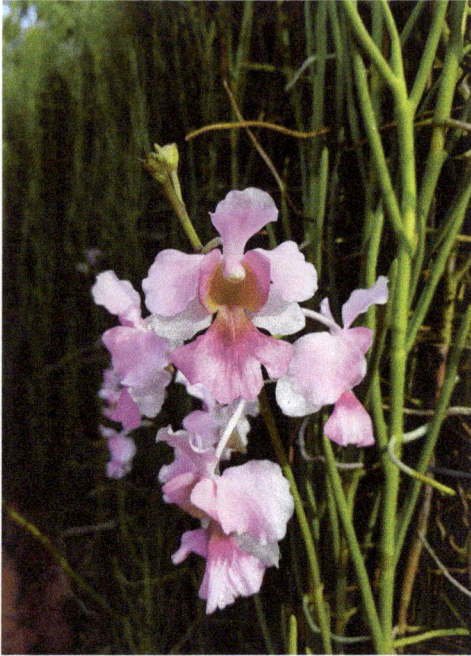

Papilionanthe Miss Joaquim, also known as *Vanda Miss Joaquim* (Photo credit: Soh Shan Bin).

plant by Agnes Joaquim and not an orchid discovered by accident.[41] This beautiful Singapore orchid, as one of the national flowers of the Commonwealth countries, was sewn on the veil worn by Meghan Markle at her royal wedding with Britain's Prince Harry.[42]

---

[41] Wright, N. *et al.* (2018).

[42] The Straits Times (2018). https://www.straitstimes.com/world/singapore-orchid-on-meghan-markles-wedding-veil-featuring-national-flowers-of-commonwealth?xtor=CS3–17. Published 19th May 2018.

Vanda, pronounced "Wanda" in Sanskrit, is derived from the Sanskrit name for *Vanda tessellate*. This is a highly popular horticultural genus for its showy *Vanda* flowers that are long-lasting and often fragrant. It is prevalent in India and throughout Southeast Asia and southern China.

Types of *Vanda* used as medicinal herbs include *Vanda coerulea* Griff. ex Lindl., *Vanda concolor* Blume, *Vanda cristata* Wall ex Lindl., *Vanda tessellate* (Roxb.) Hook. ex G. Don and *Vanda testacea* (Lindl.) Rchb.f.

Vanda coerulea Griff. ex Lindl. (Photo credit: Singapore Botanic Gardens).

7.2 Special Features of *Vanda*

Vanda plants are medium to large. Most of them are epiphytic, with sturdy stems and strap-like leaves which are glabrous, flat and arranged in two rows on a single plane, with leaves from each row stacked alternately. In the past, there was another group of *Vanda* which have terete (cylindrical or slightly tapering) leaves, but now it is known to be a separate genus known as *Papilionanthe schltr.* The inflorescences are axillary and carry multi-coloured flowers that are usually flat and have a short spur on the lip. The flowers are often fragrant. However, among the medicinal species of *Vanda*, only *Vanda tessellate* emits a scent as it contains volatile oils like linalool.[43]

7.3 Traditional Medical Uses

India: *Vanda* has a long history of use as medicine by traditional healers in India. The major traditional medicine systems used in India are Ayurveda, Siddha and Unani, among which Ayurveda is the oldest. The word Ayurveda comes from "Ayus", which means life, and "Veda," which means science, so Ayurveda literally means science of life.[44] It is a branch of Indian science that includes medicine, herbalism, taxology, anatomy, surgery and alchemy.[45] The practice of Ayurveda employs the use of plants, animals and minerals or metals as therapeutic agents.

---

[43] Teoh, E.S. (2016).

[44] Mukherjee, P.K. (2001).

[45] Rasna. https://www.wisdomlib.org/definition/rasna. Retrieved 15 Jul 2018.

Orchid species *Vanda testacea* and *Vanda tessellata* are both used in Ayurveda. The roots of these two orchids are often sold as Rasna in Ayurvedic medicine. Rasna was used as an ancient tonic for ageing. It can either be taken orally or applied externally to treat nervous disorders, rheumatism and scorpion stings. Rasna is also an important component of Astavarga, a group of eight medicinal herbs used in Ayurveda believed to have rejuvenating and health-promoting properties. They are also highly valued as tonics for the liver and brain, and are used to treat inflammation and heal fractures.

Besides its popular use in Ayurveda, the different plant parts of *Vanda tessellata* are used for various medicinal purposes in India. The whole plant (including the leaves and the roots) can be used to treat fractures. The roots of *V. tessellate* are made into ear drops and used as a female contraceptive, whereas the leaves are thought to be able to prevent wounds from developing into purulent sores and also to treat tearing and earache. Lastly, the "bark" together with other herbs are applied to sores.

Vanda is also used in the Unani medical system, which originated from Greece and was based on the teachings of Hippocrates. It was introduced by the Arabs and was brought to India by scholars and physicians of Unani medicine, who fled from Persia and central Asia when the Mongols invaded that region. In

Unani medicine, *Vanda* roots are used to treat dyspepsia, bronchitis, inflammation and cough. *Vanda tessellata* was also used as a laxative by the Unani practitioners.

The use of *Vanda* for medicinal purposes is widespread in India. Natives of Arunachal Pradesh, a state rich in orchids, use the juice made from the leaves of *Vanda coerulea* as an expectorant and also to treat eye diseases, dysentery, diarrhoea and skin diseases. The leaf juice of another species, *Vanda cristata*, which is similarly used as an expectorant, is used to treat tonsillitis, bronchitis and dry cough. It is also sometimes used as a tonic to alleviate general weakness.

The leaf and root paste of *Vanda cristata* are applied to cuts and wounds. The root paste is also used for boils and dislocated bones.[46] The Tharu community of Udham Singh Nagar in the state of Uttarakhand uses *Vanda* to treat skin diseases. It has also been observed that people in the state of Madhya Pradesh use the scent of the flower of *Vanda tessellate* to treat migraine.

Thailand: In northern Thailand, *Vanda* with other plant species have been made into poultice and used by healers of the Akha tribe to treat burns. It is also used to make infusions to clean and bathe external injuries.

---

[46] Tsering, J. *et al.* (2017).

China: *Vanda* is not nearly as popular for medicinal use in China. The Chinese use *Vanda concolor* Blume to resolve dampness and clear toxins. It is also used to treat pain due to rheumatism and sores of boils.[47]

7.4 Biomedical Analysis of Medicinal Properties of *Vanda*

Vanda tessellata, also known as *Vanda roxburghii*, has been studied for its wound-healing effects. In a pre-clinical study, rats administered with *Vanda tessellata* were found to have wounds reduced in diameter more effectively than those from the control group, which were administered only carboxymethyl cellulose (sodium salt).[48] The wound healing process includes coagulation, inflammation, granulation, fibroplasia, collagenesis, wound contraction and epithelization. The extract of *Vanda tessellata* has been found to contain two anti-inflammatory compounds, hepcosame ($C_{27}H_{56}$) and octacosonol ($C_{28}H_{58}O$), which may have prevented further damage caused by inflammation and thus helped in the healing of wounds.

The flower of *Vanda tessellata* is strongly scented due to the presence of volatile oils, with linalool accounting for 23% of the total. Linalool is present in

---

[47] *Zhong Hua Ben Cao* 中华本草 (1999), Vol. 8, pp. 759.
[48] Nayak, B.S. (2005).

many plants, and it is bacteriostatic, fungistatic, anti-inflammatory and relieves pain. This probably explains *Vanda tessellata*'s therapeutic effects in the treatment of wounds and in relieving pain. Linalool is also an anti-oxidant, which could explain the use of the orchid as an anti-ageing tonic.

Protection from UV: Plants protect themselves from excessive solar radiation by synthesising polyphenolic compounds that have anti-oxidant properties and can absorb ultraviolet (UV) radiation. Skin cells subjected to UV exposure synthesize an enzyme COX-2 which leads to an increase in production of a compound known as PGE-2. COX-2 is associated with the pathophysiology of inflammation and cancer while PGE-2 is known to play an important role in inflammation and is in abundance in aged skin. Therefore, inhibition of COX-2 and PGE-2 production are important mechanisms to protect skin exposure to UV.

A French study demonstrated that the polyphenolic compounds extracted from *Vanda coerulea* inhibited the production of COX-2 and PGE-2. The study suggests that some compounds in *Vanda coerulea* might have potential use for protecting skin against UV radiation and ageing.[49]

---

[49] Simmler, C. *et al.* (2010).

It has been pointed out that many photoprotective skin products and skin whiteners carry orchid ingredients in their patent.[50]

***Vanda* against Ageing**: There has been increasing evidence for ageing due to mitochondrial inactivity. Increases in oxidative stress and reduction in cellular energy of mitochondria occur naturally with age. Eucomic acid and one of its derivatives that is isolated from the *Vanda teres* roots have been found to maintain mitochondrial functions as well as stimulate mitochondrial biogenesis. Although there are no written records of *Vanda teres* (also known as *Papilionanthe teres* (Roxb.) Lindl.) being used traditionally as anti-ageing agents, eucomic acid and its derivative from *Vanda teres* could potentially constitute natural ingredients in remedies for age-related conditions and in anti-skin ageing agents.

---

[50] By Dr. Teoh Eng Soon at the Orchid symposium. See also Teoh, E.S. (2017) pp. 668.

Chapter 8

Cypripedium

8.1 Introduction

The name of the genus *Cypripedium* is derived from two words, "*Cyprus*" and "*pedium*". "*Pedium*" in Latin means slipper, thus *Cypripedium* has the meaning of slipper of Cyprus, a small scenic island in the Eastern Mediterranean known as Petra tou Romiou being, the birthplace of the legendary Greek goddess of love and beauty. Cyprus hosts different habitat types that reflect into a great variety of micro-environments that can be home to a considerable number of plant species. About 2000 taxa (species, subspecies, varieties) have been identified in Cyprus, out of which 144 are endemic to the island.

Cypripedium is one of the five genera of the sub-family of the Orchid family known as Cypripedioideae, or lady's slipper orchids. The *Cypripedium* genus contains 58 species. They can be found across Europe, Russia, China, Central Asia, Canada, the United States, Mexico and Central America. The other common names for *Cypripedium* include *shao lan* (杓兰), moccasin flower, camel's foot, squirrel foot, steeple cap, Venus' shoes and whippoorwill shoe.

8.2 Special Features of *Cypripedium*

The slipper orchids are unique due to their showy flowers, with the lip shaped like a pouch or slipper.

Depending on the species, the 'slipper' of the flower varies in size and can be as large as a chicken egg. The dorsal sepal looks like a hood over the lip and the petals are usually free, spreading or droopy. The stems are long and erect with leaves growing along their length for most of the species. Slipper orchids are largely terrestrial.

Traditional medical uses

In China and Taiwan, the entire plant of *Cypripedium debile*, and the roots and stems of *Cypripedium fasciolatum* and *Cypripedium macranthos,* are used to improve blood flow, reduce swelling and pain and as a diuretic. In parts of China like Sichuan, Hebei and Shanxi, *Cypripedium fasciolatum* is also used to clear phlegm and to treat generalised oedema, pain in the joints, swelling of the lower extremities, fractures and other traumatic injuries.[51] The other uses of *Cypripedium macranthos* would be to treat oliguria, leucorrhoea, gonorrhoea, rheumatism, traumatic injuries, dysentery and illness resulting from overwork.[52] When dried, flowers of the *Cypripedium macranthos* are made into powder and used to stop bleeding in wounds.

Similarly, the entire plant of *Cypripedium formosanum* is used to help improve blood flow and relieve pain. Besides that, it is believed to regulate menses, expel gas and relieve itch. The root and stem have been used to treat malaria, snake bites, traumatic injury and rheumatism.

Cypripedium henryi, another species from this genus, is found in Shanxi, Gansu, Hubei and southwest China. The roots are also used to improve *qi* and blood circulation, reduce swelling and pain, and treat conditions such as "cold in the stomach", pain in the back and lower limbs and pain due to injury.

---

[51] Teoh, E.S. (2016).

[52] *Zhong Hua Ben Cao* (1999), Vol. 8, pp. 704.

Cypripedium franchetii is used in Chinese medicine to regulate the flow of *qi*, improve blood circulation and eliminate obstruction. It is used to relieve coughs and help with chest and epigastric pain. The roots and stems of *Cypripedium franchetii* share the same herb name, *Wugongqi* (蜈蚣七), as *Cypripedium fasciolatum* and they have the same therapeutic actions.[53]

Cypripedium japonicum, also known as Japanese *Cypripedium* or Korean lady's slipper,[54] is believed to be able to regulate *qi*, improve blood flow, relieve pain, remove toxins and treat malaria. It is used to treat conditions like physical injuries, backache, rheumatism, irregular menstruation, unknown causes of swelling and pain, snake bites and skin itch.[55]

Cypripedium himalaicum is used in Chinese medicine to treat hernia, women with infertility and pain at the waist. In Nepal, it is used to treat urinary problems like difficulty in passing urine and urinary stones, as well as heart and lung diseases and coughs.[56]

---

[53] *Zhong Hua Ben Cao* (1999),Vol. 8, pp. 701.

[54] Cypripedium japonicum. https://en.wikipedia.org/wiki/Cypripedium_japonicum. Retrieved on 31 Jul 2018.

[55] *Zhong Hua Ben Cao* (1999), Vol. 8, pp. 703.

[56] Teoh, E.S. (2016).

Cypripedium himalaicum (Photo credit: Naresh Swami).

Cypripedium margaritaceum is a folk herb in Yunnan that is boiled and drunk to nourish the liver and kidneys. It moderates *qi* and blood, promotes diuresis to relieve oedema, and improves blurred vision or night blindness.[57]

In Yunnan, China *Cypripedium tibeticum* can be found and its roots are thought to contain anti-inflammatory properties and can prevent pain.

---

[57] Zhong Hua Ben Cao (1999), Vol 8, pp 703.

They can increase urine output, reduce swells and relieve pain, improve blood circulation to treat menstrual disorders. Tibetans take the concoction made from its roots to treat rheumatism, leg oedema, external injuries, gonorrhoea and leucorrhoea.

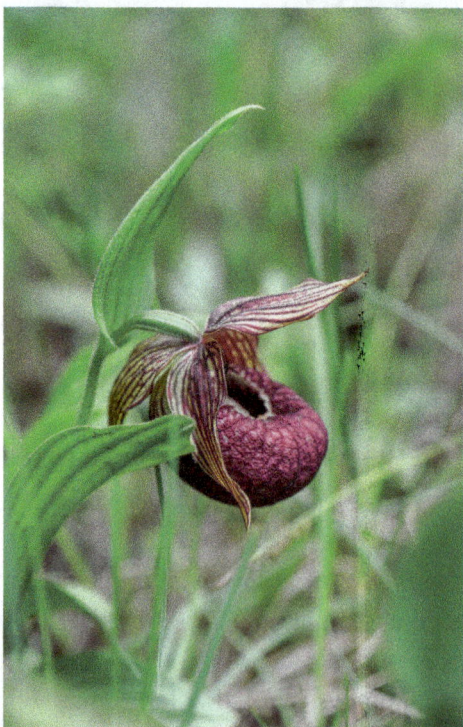

Cypripedium tibeticum (Photo credit: Jiang Tianmu).

In India, the roots of *Cypripedium calceolus* are made into powder and added into sugar water to act as a sedative, promote sleep and reduce pain. A tea prepared from its roots has a calming effect and is given to people who are nervous, irritable or having headaches.

Indians use the roots of *Cypripedium elegans* and *Cypripedium pubescens*. They are known to have anti-spasmodic, diaphoretic, hypnotic, nervine and sedative effects. They are used to treat disorders of the nervous system and as a nerve tonic for cases like hysteria or madness, spasm, fits, epilepsy and rheumatism.

Cypripedium pubescens is also used in the treatment of diabetes, dysentery, paralysis, convalescence, impotence and malnutrition.[58] *Cypripedium pubescens* is one of the most valuable medicinal herbs in America and is used in a similar manner as in India. It is quite extensively used to relieve pain and is preferable to opium as it does not have narcotic effects. It is also used in the treatment of worms in the stomach.

Cypripedium parviflorum is one of the most important medicinal lady's slipper in North America. The dried rhizomes are made into powder and mixed in sugar water to treat insomnia, anxiety, fever, headache, neuralgia, emotional tension, palpitations, tremors, irritable bowel syndrome, delirium, convulsions

---

[58] Hossain, M.M. (2010).

due to fever and to ease the pain of menstruation and childbirth.

The roots of *Cypripedium cordigerum* are used as a tonic in Nepal and the roots and leaves of *Cypripedium guttatum* are used in eastern Russia and Siberia to treat epilepsy. Biomedical explanations for the medical properties of this genus are lacking. It is known, however, that lusianthrin and chrysin were isolated from seedlings of *C. macranthos* that had developed shoots. Lusianthrin and chrysin are anti-fungal compounds.

Chapter 9

Habenaria

9.1 Introduction

Habenaria, also known as the rein orchids or the bog orchids, is a large genus of orchids that is widely distributed in both tropical and subtropical regions, and can be found in every continent except Antartica. The orchids are mainly terrestrial and the plants are short, with leaves arranged in a rosette at the base of the stem or up the stem. The plant has fleshy tubers and succulent stems, and inflorescences are unbranched. Flowers are usually small, white, green or yellow in colour.

This orchid is unique because of its large lip that is flat and multi-lobed, and classification of species in this genus is mainly based on the lip form and petal shape.[59]

Habenaris ciliolaris (Photo credit: 蘇清記 Su Qing Ji).
Source: OrchidRoots, http://bluenanta.com/orchid/

---

[59] Teoh, E.S. (2016).

Habenaria thrives in areas with distinct wet and dry seasons. It goes through a dormant phase before flowering. With the arrival of the rain, an aerial shoot emerges and grows rapidly during the heavy rainfall period and flowers. The aerial portion dies down after the heavy rains stop. The underground tuber that remains grows and sends out new aerial shoots when the rains come again.[60]

The generic name for this orchid genus is derived from the Latin word *habena* (bridle, whip, strap, veins), which describes the thread-like fringe on the lip in some species, e.g. *Habenaris ciliolaris*.[61]

9.2 Traditional Medical Uses

Habenaria are used extensively in China and India as medicine. Interestingly, however, the species used in each country for medicinal purposes are different, except in the case of *Habenaria arietina*, which is used as a tonic in both countries.

China: *Habenaria* species used by the Chinese include herbs like *Habenaria davidii* and *Habenaria dentata* with the therapeutic action to improve kidney functions. Hence, the word *shen* (肾) is incorporated in the name of the herbs, e.g. shuangshencao (双肾草) and

---

[60] Abraham, A. *et al.* (1981).

[61] Teoh, E.S. (2016), pp. 441.

shuangshenzi (双肾子).[62] *Shen*, or kidney, in Chinese medicine represents a set of functions that includes urine excretion, growth, reproduction, marrow and brain functions and warming of the internal organs.

Both *Habenaria aitchisonii* and *Habenaria delavayi* are collected in Yunnan, Sichuan and Tibet, and can be used to treat nephritis. *Habenaria delavayi*, known in Chinese medicine as *Jishenshen* (鸡肾参), is also used as a tonic to strengthen kidneys for improving symptoms associated with kidney weakness, like lower back pain, dizziness and tinnitus.

Besides treating kidney-associated problems like nocturnal emissions, impotence, urinary problems and leucorrhoea, the Chinese use *Habenaria ciliolaris* to treat gonorrhoea, stomachache, tuberculous cough, kidney infection and snake bites. The tincture of *Habenaria ciliolaris* is taken in rice wine consumed with rice as a tonic for internal injuries.

Habenaria dentata, Habenaria linguella, Habenaria pectinate and *Habenaria petelotii* are used to treat lung conditions. *Habenaria dentata* is used to treat coughs caused by tuberculosis, whilst *Habenaria linguella* is used to clear heat in the lungs. *Habenaria pectinata* has been used to treat coughs arising from general body debility, and *Habenaria petelotii* for coughs from heat in the lungs.

---

[62] *Zhong Hua Ben Cao* (1999), Vol. 8, pp. 727–731.

The Chinese use *Habenaria aitchisonii, Habenaria delavayi, Habenaria ciliolaris* and *Habenaria davidii* to treat hernia.

India: *Habenaria arietina* and *Habenaria edgeworthii* are given the Sanskrit name *Riddhi* and *Vriddhi* respectively. They are known as tonics for the treatment of fainting spells, and are also used for deworming and as a blood purifier.

The name *Riddhi* is also used collectively for those particular *Habenaria* species used in Ayurveda for medicinal purposes such as a revitalizing tonic or as an aphrodisiac. *Habenaria arietina*, also known as *Habenaria intermedia*, together with *Habenaria edgeworthii* are important ingredients of the *Ashtavarga* group of herbs. The *Ashtavarga* is a formula with eight herbs, containing *Habenaria* as well as two other orchids, *Malaxis muscifera* and *Crepidium acuminatum*. The *Ashtavarga* herbs are used to make a cooked mixture known as *Chyawanprash* that is consumed as a dietary supplement to help fight coughs and colds.[63] There is no clear distinction between *Riddhi* and *Vriddhi* sold in the market. The larger tubers of the *Habenaria* are used as *Vriddhi* and the smaller ones as *Riddhi*.

---

[63] Chyawanprash. https://en.wikipedia.org/wiki/Chyawanprash. Retrieved 31 Aug 2018.

Chyawanprash is popular in Nepal as well. The Nepalese use it to treat an amazing variety of ailments including thirst, fever, coughs, asthma, anorexia, haematemesis, worms, emaciation, general debility, skin diseases, leprosy, cataplexy and insanity. There is also popular belief that *Chyawanprash* can help to boost intelligence.

Habenaria commelinifolia together with *Saraca indica* (Ashoka tree) are believed to cure spermatorrhoea in India. These two herbs are boiled together in 1000 ml of water until the concoction is only left with 100 ml, and patients are instructed to take six to eight drops of this medication for ten days on an empty stomach. This plant is also eaten as a vegetable and claimed to be a blood purifier that cures blebs (small blisters) on the palm.

Many *Habenaria* species are used to treat various types of bites and stings caused by snakes, scorpions and insects. *Habenaria diphylla* is used for treating insect bites in Thailand as well. In India, a paste made from *Habenaria furcifera* and *Habenaria roxburghii* is used for snake bites, while *Habenaria hollandiana* is used to treat scorpion stings and maggot-infected sores.

In the Ayurveda system, *Habenaria furcifera, Habenaria plantaginea* and *Habenaria roxburghii* have very similar therapeutic actions. Their tubers are used by the traditional medical practitioners to treat wasting diseases, fever, blood disorders and fainting.

Habenaria plantaginea together with other ingredients are commonly used in India and Bangladesh for chest pain and stomachaches.

Habenaria longicorniculata was reported during an India folk practitioners' meeting, known as *Natti Vaidyas Sammelan*, that its fresh tubers were eaten to reduce scrotal enlargement. The whole plant is also used to help with pain and swelling. Its tuber mixed with the same volume of turmeric power and made into a paste is claimed to treat leukoderma, a skin disorder where the skin has white patches.[64]

---

[64] Teoh, E.S. (2016), pp. 450.

Bibliography[‡]

Abraham, A. *et al.* (1981). *Introduction to Orchids, with Illustrations and Descriptions of 150 South Indian Orchids*. TPGRI, Trivandrum.

Bulpitt, C.J. (2005). The uses and misuses of orchids in medicine. *Quart. J. Med.* 98(9): 625–631, https://doi.org/10.1093/qjmed/hci094

Burkill, I.H. (1935). (1966 reprint, 2nd ed., with contributions by Birtwistle, W., Foxworthy, F.W., Scrivenor, J.B. and Watson, I.G) *A Dictionary of Economic Products of the Malay Peninsula, Vol. II*. London: Crown Agents for the Colonies, Kuala Lumpur.

[‡]Additional references cited in Chapters 4 and 6 may be found at the end of the respective chapters.

Chen, S.C. *et al.* (1982). A general review of the orchid flora of China. In: Arditti, J. (ed.) *Orchid Biology: Reviews and Perspectives II*. Cornell, Ithaca.

Chyawanprash. https://en.wikipedia.org/wiki/Chyawanprash. Retrieved 31 Aug 2018.

Cypripedium japonicum. https://en.wikipedia.org/wiki/Cypripedium_japonicum. Retrieved on 31 Jul 2018.

Dendrobium. https://en.wikipedia.org/wiki/Dendrobium. Retrieved 23 Jun 2018.

Deng, Y. *et al.* (2018). Chemical characterization and immunomodulatory activity of acetylated polysaccharides from *Dendrobium devonianum*, *Carbohydr. Polym.* 180: 238–245.

Nguyen, V.D. (1993). *Medicinal Plants of Vietnam, Cambodia and Laos*. World Health Organization, Manila.

He, X. *et al.* (2017). Bletilla striata: Medicinal uses, phytochemistry and pharmacological activities. *J. Ethnopharmacol.* 195: 20–38.

Hew, C.S. *et al.* (2006). Orchids in Chinese medicine. *Innovation* 6(2): Schools Section.

Hossain, M.M. (2010). Therapeutic orchids: Traditional uses and recent advances — An overview. *Fitoterapia* 82(2): 102–140.

Lawler, L.J. (1986). Orchid ethnobotany in the Asean Area. In: Rao, A.N. (ed.) *Proceedings of the 5th Asean Orchid Congress*. Parks & Recreation Department, Ministry of National Development, Singapore, pp. 42–45.

Leung, P.C. (2017). Research on two wonderful herbs of orchid family for aging problems, presented at *Orchid*

Symposium: From Fundamental Research to Medical Applications. Singapore, 8 Nov 2017.

Li, R. *et al.* (2017). Anti-influenza A virus activity of Dendrobine and its mechanism of action. *J. Agric. Food Chem.* 65(18): 3665–3674.

Li, S.P. (2017). Study on polysaccharides from Dendrobium in China, presented at *Orchid Symposium: From Fundamental Research to Medical Applications*. Singapore, 8 Nov 2017.

Magwere, T. (2009). Escaping immune surveillance in cancer: Is denbinobin the panacea? *Br. J. Pharmacol.* 157(7): 1172–1174.

Maritim, A.C. *et al.* (2003). Diabetes, oxidative stress, and antioxidants: A review. *J. Biochem. Mol. Toxicol.* 17(1): 24–38.

Meng, L.Z. *et al.* (2013). Effects of polysaccharides from different species of Dendrobium (Shihu) on macrophage function. *Molecules* 18(5): 5779–5791.

Metabolomics/Metabolites. https://en.wikibooks.org/wiki/Metabolomics/Metabolites. Retrieved 19 Feb 2019.

Mukherjee, P.K. (2001). Evaluation of Indian traditional medicine. *Drug Inf. J.* 35: 623.

Nayak, B.S. (2005). Evaluation of wound healing activity of *Vanda roxburghii* R. Br (Orchidacea): A preclinical study in a rat model. *Int. J. Lower Extrem. Wounds* 4(4): 200–204.

Orchids in folklore and mythology (11 Aug 2015). https://www.houzz.com/discussions/3291692/orchids-in-folklore-and-mythology. Retrieved 2 Feb 2019.

Perry, L.M. *et al.* (1980). *Medicinal Plants of East and Southeast Asia: Attributed Properties and Uses.* MIT Press, Cambridge, MA.

Rasna. https://www.wisdomlib.org/definition/rasna. Retrieved 15 Jul 2018.

Sheehan, T. & Sheehan, M. (2009). *An Illustrated Survey of Orchid Genera.* Timber Press.

Shi, Y. *et al.* (2017). Antiviral activity of phenanthrenes from the medicinal plant *Bletilla striata* against influenza A virus. *BMC Complement. Altern. Med.* 17(1): 273.

Simmler, C. *et al.* (2010). Antioxidant biomarkers from *Vanda coerulea* stems reduce irradiated HaCaT PGE-2 production as a result of COX-2 inhibition. *PLoS One* 5(10): e13713.

Teoh, E.S. (2016). *Medicinal Orchids of Asia.* Springer, Switzerland.

Teoh, E.S. (2017). Exploring orchids as medicine, presented at *Orchid Symposium: From Fundamental Research to Medical Applications.* Singapore, 8 Nov 2017.

The orchid flower, its meanings and symbolism. http://www.flowermeaning.com/orchid-flower-meaning/ Retrieved 2 Feb 2019.

The Straits Times (2018). https://www.straitstimes.com/world/singapore-orchid-on-meghan-markles-wedding-veil-featuring-national-flowers-of-commonwealth?xtor=CS3-17. Published 19 May 2018.

Tsering, J. *et al.* (2017). Medicinal orchids of Arunachal Pradesh: A review. *Bull. Arunachal Forest Res.* 32(1&2): 1–16.

Wang, C. *et al.* (2006). A polysaccharide isolated from the medicinal herb *Bletilla striata* induces endothelial cells proliferation and vascular endothelial growth factor expression *in vitro*. *Biotechnol. Lett.* 28(8): 539–543.

魏等 (2008). 金叉石斛提取物抗白内障的体外实验研究. 现代中药研究与实践. 22(2): 27–31.

Wright, N. *et al.* (2018). Blooming lies: The Vanda Miss Joaquim story. *Biblioasia* 14(1): 2–9.

Xu, J. *et al.* (2010). Fast determination of five components of coumarin, alkaloids and bibenzyls in *Dendrobium* spp. using pressurized liquid extraction and ultra-performance liquid chromatography. *J. Sep. Sci.* 33(11): 1580–1586.

Yang, Y.J. *et al* (2009). Effect of scoparone on dopamine biosynthesis and L-DOPA-induced cytotoxicity in PC12 cells. *J. Neurosci. Res.* 87(8): 1929–1937.

Zheng, C. *et al.* (1998). Bletilla striata as a vascular embolizing agent in interventional treatment of primary hepatic carcinoma. *Chin. Med. J.* 111: 10060–11063.

Zhong Hua Ben Cao 中华本草 (1999). Shanghai Scientific & Technical Publishers, Shanghai.

Index